ADVANCED SUBMARINE TECHNOLOGY AND ANTISUBMARINE WARFARE

HEARING

BEFORE THE

SEAPOWER AND STRATEGIC AND CRITICAL MATERIALS SUBCOMMITTEE

AND THE

RESEARCH AND DEVELOPMENT SUBCOMMITTEE

OF THE

COMMITTEE ON ARMED SERVICES HOUSE OF REPRESENTATIVES

ONE HUNDRED FIRST CONGRESS

University Press of the Pacific
Honolulu, Hawaii

Advanced Submarine Technology and
Antisubmarine Warfare

by
U.S. House of Representatives

ISBN: 1-4102-2431-7

Copyright © 2005 by University Press of the Pacific

Reprinted from the 1990 edition

University Press of the Pacific
Honolulu, Hawaii
http://www.universitypressofthepacific.com

ADVANCED SUBMARINE TECHNOLOGY AND ANTISUBMARINE WARFARE

SEAPOWER AND STRATEGIC AND CRITICAL MATERIALS SUBCOMMITTEE

CHARLES E. BENNETT, Florida, *Chairman*

THOMAS M. FOGLIETTA, Pennsylvania	FLOYD SPENCE, South Carolina
NORMAN SISISKY, Virginia	DUNCAN HUNTER, California
SOLOMON P. ORTIZ, Texas	HERBERT H. BATEMAN, Virginia
GEORGE J. HOCHBRUECKNER, New York	CURT WELDON, Pennsylvania
JOSEPH E. BRENNAN, Maine	JOHN G. ROWLAND, Connecticut
OWEN B. PICKETT, Virginia	BEN BLAZ, Guam
EARL HUTTO, Florida	

ROBERT E. SCHAFER, *Professional Staff Member*
LAWRENCE J. CAVAIOLA, *Professional Staff Member*
DIANE L. HARVEY, *Staff Assistant*

―――――

RESEARCH AND DEVELOPMENT SUBCOMMITTEE

RONALD V. DELLUMS, California, *Chairman*

DAVE McCURDY, Oklahoma	WILLIAM L. DICKINSON, Alabama
THOMAS M. FOGLIETTA, Pennsylvania	ROBERT W. DAVIS, Michigan
DENNIS M. HERTEL, Michigan	BOB STUMP, Arizona
GEORGE (BUDDY) DARDEN, Georgia	DUNCAN HUNTER, California
GEORGE J. HOCHBRUECKNER, New York	CURT WELDON, Pennsylvania
OWEN B. PICKETT, Virginia	JON KYL, Arizona
MICHAEL R. McNULTY, New York	ROBERT K. DORNAN, California
CHARLES E. BENNETT, Florida	
PATRICIA SCHROEDER, Colorado	
BEVERLY B. BYRON, Maryland	

CARL T. BAYER, *Professional Staff Member*
MARILYN A. ELROD, *Subcommittee Professional Staff Member*
THOMAS S. HAHN, *Counsel*
MARY E. VESELY, *Staff Assistant*

(II)

CONTENTS

(III)

ADVANCED SUBMARINE TECHNOLOGY AND ANTISUBMARINE WARFARE

House of Representatives, Committee on Armed Services, Seapower and Strategic and Critical Materials Subcommittee, and the Research and Development Subcommittee, *Washington, DC, Tuesday, April 18, 1989.*

The subcommittees met, pursuant to notice, at 2 p.m., in room 2118, Rayburn House Office Building, Hon. Charles E. Bennett (chairman of the Seapower and Strategic and Critical Materials Subcommittee) presiding.

OPENING STATEMENT OF HON. CHARLES E. BENNETT, A REPRESENTATIVE FROM FLORIDA, CHAIRMAN, SEAPOWER AND STRATEGIC AND CRITICAL MATERIALS SUBCOMMITTEE

Mr. Bennett. The subcommittees will come to order.

Well, if we need to go into executive session, we will have to have a quorum for that purpose, and since both committees are here, that takes 15 people. That is not likely with everything that is happening on the floor.

So at that point, instead of going into executive session, we will go into a briefing and adjourn the hearing.

Today the Seapower Subcommittee and the Research and Development Subcommittee meet in joint session to continue their examination of advanced submarine technology and antisubmarine warfare.

The hearing will begin in open session, but will probably have to close to receive national security information. When we obtain a quorum of the joint subcommittees, we will have a vote on a motion to close the hearing.

On March 21, the joint subcommittees received testimony from a panel of experts convened by the committee Chairman, Mr. Aspin, to review the status of our Nation's submarine technology and antisubmarine warfare efforts.

The panel made a number of important findings and recommendations. Among the most important were the following: First, that failure to deal adequately with quiet submarines under the control of our adversaries could have a profound effect on our national security; and, second, that the Navy needed to improve its strategic vision for dealing with ASW as a prerequisite to developing the ASW system needed for the future.

So the purpose of today's hearing is to hear from the Navy on its views about submarine technology, ASW, and the panel's report. In particular, the subcommittees are interested in the process by which the Navy determines its allocation of resources to ASW, whether that process is adequate to meeting the ASW challenge of the future, and whether we can count on that process to tell us if we need more money for ASW.

In addition, because of the importance accorded to them by the expert panel, the current and proposed programs of the Defense Advanced Research Projects Agency in both submarine technology and ASW research will be examined here today.

Testifying today for the Navy will be Vice Adm. William D. Smith, the Deputy Chief of Naval Operations for Navy Program Planning. Admiral Smith is accompanied by representatives from the Office of the Secretary of the Navy, the Naval Sea Systems Command, and the Office of the Deputy Chief of Naval Operations for Naval Warfare.

Testifying from the Defense Advanced Research Projects Agency will be Mr. Robert Moore, Deputy Director for Systems and Technology.

I must say that my memory is that over a year ago, Mr. Carlucci, referring to the highly critical report of the Defense Science Board, directed the Secretary of the Navy and the CNO to present him on May 1 a well-thought-out plan by which the Navy would organize, integrate, not only technology transition but the entire ASW mission at the highest level.

This committee doesn't want to micromanage anything, but we have not yet received as much information as we would like to on this subject matter, and we hope today we will have a feeling of urgency and that there is in place a procedure that will lead us to answers to some of these problems.

Before calling on the panel for their testimony, Mr. Spence, do you have anything you would like to say at this time?

Mr. SPENCE. Nothing, except that I am concerned about this report of this panel, as you are. I am anxious to find out from those of you who are in a better position to tell me just what the true situation is, and I welcome you today.

Thank you very much.

Mr. BENNETT. Thank you.

You can count on this committee wanting to find the answers and to carry them out. We do want to meet with the urgent action on the part of Congress, whatever urgent action is required by what the Navy has to tell us.

Mr. BENNETT. We will now proceed with Admiral Smith's testimony.

STATEMENT OF VICE ADM. WILLIAM D. SMITH, U.S. NAVY, DEPUTY CHIEF OF NAVAL OPERATIONS (NAVY PROGRAM PLANNING)

Admiral SMITH. Good afternoon, Mr. Chairman. The Navy very much appreciates the opportunity to come before your subcommittee and give our view of the current ASW progress that the Navy is making, both in the integrated aspects and in the individual programs that may be of interest to your members.

Mr. BENNETT. Is it urgent that you have a czar?

Admiral SMITH. I think the CNO would say that he is the czar of ASW. He certainly has the background, the interest, the focus and the intent, and will remain in charge of that part of the Navy's vital war-fighting capability.

We have for a number of years made ASW our number one war-fighting priority. War fighting isn't our number one priority for just that purpose, but there is also a high priority on maintaining and retaining our skilled personnel.

War fighting is not meant as a qualifier to reduce the importance we put on ASW.

Mr. BENNETT. We have had civilians who have had this responsibility—I am not choosing between civilians and people in uniform, but we did have a feeling of being impelled in the right vigorous position by civilian leadership which is no longer in the Navy.

Is there a Navy man other than those in uniform who is carrying the ball on this?

Admiral SMITH. That is correct. The only cross-platform integrated approach is represented in a uniformed side of the Navy. We have four Assistant Secretaries of the Navy that work in the area of research and engineering, in shipbuilding and logistics, in the financial management world, and in the manpower and reserve affairs world.

So the only overview of this, although both Secretary Ball and Under Secretary Garrett are very interested, very keenly aware and very much up to speed on where the Navy is going in ASW and have given the CNO essential support that he needs to keep this at the top of our level of visibility.

The Navy has done a number of things in the last few years to sharpen that focus. In 1987, our Director of Naval Warfare, who is the requirements expert for the Chief of Naval Operations, was divested of all of his individual programs, not that he had that many, but he had programs which caused him to occasionally see a need to support the programs that were directly under his resource sponsorship at the expense of an overall view of war fighting across the various mission areas of the Navy.

So he was divested of those responsibilities in 1987 to give him a better opportunity to be an honest umpire for the entire process. In addition to that step, based on the Defense Science Board report that you referred to, the Secretary of the Navy, with the recommendation of the CNA, established the Office of ASW Technical Development, and Rear Adm. Dan Wolkensdorfer is the new Admiral in charge of that position.

That is exploratory in basic research that he will focus on; some advanced development, but principally from the exploratory research and basic research areas.

Additionally, with this budget submit, the 1990 budget, the Navy sent forward a plan to highlight, identify and use an R&D submarine, something we have not done in a number of years, with a first line platform.

The U.S.S. *Memphis,* a 688 class submarine, will start this August working R&D missions as its fundamental priority. It will maintain its war-fighting capability and be able and trained to deploy, but will spend the bulk of its time working in R&D.

Additionally, in the transition from basic and exploratory research into systems, we have established a line in the budget which will be managed by Rear Admiral Evans from NAVSEA who will take the transition programs that DARPA, in its basic research exploratory development areas, has fielded, which the Navy feels there is a good requirement for, which there is an immediate need for, and he will manage the transition of those into the Navy submarine systems.

We feel those initiatives which are certainly not all we have done but represent the organizational focus have been done to retain the edge we believe we have in the world of ASW.

The submarine issue will be discussed in response to your questions, and we hope to be able to demonstrate to you also that the SSN-21 is on the leading edge of the available technology at the time we have to start bending metal.

With that, I would defer for questions or for the remarks by Mr. Robert Moore. It is your choice.

Mr. BENNETT. I think we will hear from the other members of the panel before we get to questions.

PREPARED STATEMENT OF VICE ADM. WILLIAM D. SMITH

NAVY POSTURE STATEMENT ON ANTISUBMARINE WARFARE

Many reports say advances in the Soviet submarine quieting
program threaten to challenge our capability to maintain the ASW
advantage. Reports, fleet performance, and Congressional hearings show also the Navy has recognized fully over the last three
decades the importance of maintaining our ASW advantage. In
responding to the most recent in the series of submarine quieting
advances, the CNO initiated as early as 1983 a number of steps
which will maintain our current superiority. Further, ASW
programs in Washington and tactics in both the Atlantic and
Pacific Fleets are being worked very hard to ensure we maintain
our capability to lead the future threat.

PLANNING BACKGROUND

In 1984 an ASW Steering Group representing the top ASW
talent from the fleet, laboratories, industry and academia,
developed an ASW Master Warfighting Strategy and detailed the R&D
requirements that would provide the capabilities needed to carry
our current ASW superiority into the next century. In early 1985
the recommendations of the group were approved by the Chief of
Naval Operations and the Secretary of the Navy. The first ASW
Master Plan soon followed and this team plan concept was adopted
for all Naval Warfare areas. These Master Plans have served each

budget cycle well. Recognizing this process is evolutionary, the
CY 89 ASW Master Plan, provides an even more forward look. The
plan outlines and recommends those plans and programs that will
permit us to maintain our lead in ASW. The current guidance
given the Department of the Navy reflects the commitment of both
the Secretary of the Navy and the CNO to meet the serious challenge posed by the modern submarine and its future descendants.
This commitment is being backed by a program submission in which
ASW funding continues to trend upward in spite of the cut in
overall spending.

RESPONSE TO SOVIET ADVANCES

Our response to Soviet advances has been broad and comprehensive. Actions have spanned the spectrum of ASW from
research and development through fleet training and tactical
employment initiatives. Our technological research has taken
advantage of all Soviet submarine vulnerabilities including
non-traditional acoustics, active and passive sonar upgrades to
existing systems, an aggressive program in low frequency active
sonars and improved weapons lethality. We also are investigating
potential non-acoustic methods and the fleets have developed
tactical initiatives, programs and adjusted their training to
address quieter submarines.

These efforts have led to such systems as the AN/SQQ-89

system for surface combatants; the AN/BSY-1 combat control system
in the improved SSN-688 class submarines; the Mark 50 and Mark 48
ADCAP torpedoes and a new standoff ASW weapon; the SSN-21, and
associated advanced combat system, the BSY-2; the P3C replace-
ment, LRAACA, with its new update IV combat system; the S-3
Weapon System Improvement Program; the SH-60F dipping SONAR helo;
and the new family of sonobuoys that complement these new airborne systems. These systems will be capable of dealing with the
best Soviet submarines of today and with continued support for
the designed product improvements, those of tomorrow. The less
capable submarines that will form a significant portion of the
Soviet inventory into the next century will be even more vulnerable. More importantly, however, most of these new systems
can be further improved via a planned backfit with new technologies that will enhance their effectiveness, even in the face
of continued Soviet quieting. These software intensive systems
can accommodate progressive improvements as we develop new
weapons and sensor technologies to pace the evolving threat.

Largely as a result of these efforts, Navy R&D devoted to
ASW has shown steady growth since Fiscal Year 1986, even though
the Navy's top line has declined in constant dollars since
fiscal year 1985. The R&D funds requested for ASW research in
the fiscal 1990 budget are approximately 35% more than the
average annual ASW R&D expenditures of the past two years. This
growth continues through 1991. Since overall Navy R&D does not
display real growth during this period, it is also important to

note that the commitment to basic ASW R&D as represented in the basic research (6.1), exploratory (6.2), and advanced technology development (6.3A) programs has remained stable. We have not reduced this element of ASW R&D although the overall ASW R&D funds have grown within a generally flat topline for R&D. In the FY90 budget, ASW R&D is funded higher than any other warfare area.

The next generation of candidate technologies is being explored now. The Navy is anxious to get to the next page in the current chapter of modern ASW applications. These are broadly based and cover a variety of new approaches that I'll be glad to discuss more fully in closed session. Significantly, several of these are relatively immune to quieting efforts. The Navy is working closely with DARPA and other agencies to exploit the most promising research, and recently has made the decision to dedicate a 688 class SSN for R&D on a continuous basis. This SSN provides a platform for rapid prototyping of new, promising technology while maintaining its warfighting capabilities. The submarine tied predominatly to the R&D program is fully funded through FY-91 and is one of Navy's high priority programs. Similarly, we have an East Coast frigate dedicated to the development of tactical low frequency SONAR and a Pacific Fleet Destroyer Squadron whose mission is the development of coordinated ASW tactics against the modern submarine. These efforts will provide the needed technical and tactical information to reinforce decisions on future directions.

RESOURCE INVESTMENT

Even as we look to the future, it is important to recognize the level of effort and extent of force structure already devoted to ASW. In the last three fiscal years (FY 1987-1989), we have spent an average of $4.6 billion a year on ASW systems and support, not including platforms. In addition, no fewer than 43 percent of our active ships (including SSNs), and over 20 percent of our active aircraft, have ASW as their primary mission area. The new generation systems noted above will make many of these platforms more effective even against quieter Soviet nuclear submarines and modern conventional submarines. We recognize the critical importance of ASW and the facts reflect it. ASW is our number one warfighting priority.

The Navy's efforts this last decade in ASW have been aggressive, have been prudent, and have been on the mark. We have sustained a commitment to improving ASW readiness. Together with our future ASW systems, we are addressing the architectures to use them and their successors effectively in a coordinated ASW environment. The Soviet submarine force is large, quiet and formidable, but the results of a recent ASW technology net assessment reveal positive indications that our programs are working. Conducted by the Director of Naval Intelligence at the direction of the House Appropriations Committee, this classified assessment was a qualitative review of all U. S. programs versus the qualitative assessment of projected Soviet ASW and submarine

capabilities in the year 2000. The assessment concluded that
"the U. S. has qualitative superiority in submarine platforms in
the year 2000". In ASW forces the assessment also concluded that
the U. S. will retain a significant lead in ASW detection and
prosecution and would have an overall advantage in submarine and
ASW capability in the year 2000.

While this net assessment is favorable for today, the ASW of
the future will require much closer coordination, with attendant
command and control architectures, requirements for environmental
data, water management, and information fusion. Plans are being
studied, worked, discussed and molded through the appraisal and
master plan process, under the direction of the DCNO for Naval
Warfare. For example, while the ASW assessment team is reviewing
the C3I requirements for this mission area, another Team is
looking at the C3I requirements for the Navy as a whole.

ORGANIZATION AND MANAGEMENT

Our efforts will be managed well. The needed leadership is
in place. There is an "ASW Czar" within the Navy, -- Admiral
C.A.H. Trost the CNO. He exercises his responsibilities across
platform lines through the Deputy Chief of Naval Operations for
Naval Warfare, who establishes the warfare requirements, and
through the DCNO for Program Planning, who ensures that the
resources are allocated in accordance with the requirements. The

Deputy Chief of Naval Operations for Naval Warfare, (OP-07) is charged with the responsibility to coordinate the development of requirements for each warfare area and to recommend program priorities and trade-offs within and among each warfare element. He reviews warfare programming proposals and recommends approval to the CNO. These actions take place in close coordination with the surface, air, and submarine platform sponsors. ASW crosses all platform boundaries and requires coordinated effort in the field and in the acquisition process to meet the difficult challenge of this unique warfare environment.

OP-07 works closely with the Deputy Chief of Naval Operations for Program Planning (OP-08). I am charged with the centralized supervision and coordination of the Navy program planning and study effort. I ensure the functions of planning, programming, budgeting and appraisal are coordinated within the Department of the Navy. Together VADM Paul Miller and I coordinate, on a daily basis, to ensure that warfighting requirements are balanced with fiscal realities in the preparation of our investment strategy.

Our role in ASW is clear as we prepare for the FY90 budget cycle, we see the rate of investment in our ASW accounts is forecast to significantly outstrip the rate of growth in the Navy's budget as a whole. The Chief of Naval Operations directed this action in making ASW our number one warfighting priority.

As noted both by recent Congressional studies and by the Defense Science Board, one area that will need particular atten-

tion is the focused management of basic R&D. This will be
especially important as we transition the next generation technologies into fleet systems. To provide this focus, the Navy has
established a "Director of ASW Development" who will address
these critical issues. He is Rear Admiral Dan Wolkensdorfer, and
I am pleased to introduce him today. He started the ASW steering
group, Team Alfa, in the 1984 time frame so he is no stranger to
ASW requirements. Also here today is RADM Tom Evans, the Director of Advanced Submarine Research and Development and RADM Jim
Fitzgerald, Director of Antisubmarine Warfare and the Chairman of
Integrated ASW Assessment Team (Team Alfa).

CONCLUSION

With the sustained support of the Secretary of the Navy,
Secretary of Defense and the Congress our ASW efforts to date
have borne fruit and, although we have good ideas on where to go,
the years ahead will be challenging. The 1992 and beyond period
will be an especially challenging one, as funding will be required for those technologies that are sufficiently mature to
begin transition to full scale development. At the same time,
the technology base will need to be sustained as noted in the
recent HASC advisory panel report. The management of a focused
ASW Tech Base will be one of Rear Admiral Wolkensdorfer's key
challenges as we begin working on the 1992 program. We concur

with the conclusion of the Defense Science Board, the HASC
Advisory Panel, and the Seapower and R&D Subcommittees that ASW
is a National priority.

In summary, the Seapower subcommittee has followed closely
our plans to meet the Soviet Submarine quieting challenge. A
road map and catalogue to ensure our current and future superiority is contained in the Master Plan. In that plan, ASW is
viewed as a total system, encompassing ASW forces and the supporting infrastructure of global ocean surveillance and intelligence, command and control, communications, and logistics. A
pervasive and comprehensive interaction of all components is
necessary to provide a solution to the Strategic ASW problem.
Navy recognizes the need and will fully address the plans and
programs required to support the Maritime and National Stra-
tegies. We have the world's finest Navy, with the world's finest
mariners. All are dedicated to insuring that we maintain the
needed technological and tactical advantages across the full
spectrum of Naval Warfare.

Mr. BENNETT. Mr. Moore.

STATEMENT OF ROBERT A. MOORE, DEPUTY DIRECTOR, SYSTEMS AND TECHNOLOGY, DEFENSE ADVANCED RESEARCH PROJECTS AGENCY

Mr. MOORE. Thank you, Mr. Chairman.

I would like to submit the full statement for the record. This statement is an unclassified version of two classified reports submitted by DARPA to the Congress as required by the 100th Congress covering our advanced submarine technology program, long-range plan, and another covering our high-payoff antisubmarine warfare technology initiative.

Both reports are dated February 1989, and I refer you to those for the classified information on the programs.

I would now like to state some key points that I believe bear on the Soviet submarine threat which, after all, is why we are here, and cover as the essence of our advanced submarine technology program what we are trying to achieve, what we think we will accomplish in that program, and the advanced ASW technology initiative that we have designed and submitted to Congress.

I have prepared these remarks for executive session. I will cover them the best I can in this open session.

I think we find ourselves, all of us in this room, in an extraordinary situation. Congress has assigned the advanced submarine technology program to DARPA. I think major portions of that program from my perspective probably should be in the Navy.

The committee has established a package of 10 distinguished experts to advise the committee on submarine and antisubmarine technology. That committee started its work in December 1987.

The report was submitted in January, and there was a hearing before you on the 21st of March. Earlier, Congress had established specific activity at the CIA in ASW. I would suggest that the advances leading to today's hearing are dominated more by Soviet actions than by the lack of U.S. actions.

The Soviet submarine program is incredibly bold, particularly risky, operationally unsafe by virtue of that; it is very extensive. Let's look at the Soviet attack submarines that have been deployed since our SSN688 had its IOC in 1976.

Since that time, the Soviets have introduced five new classes of attack submarines plus two new significant upgrades in the older submarines. I will run through that list.

The VICTOR 3 class SSN in 1979; the ALPHA class in 1978; the OSCAR class SSN in 1981; the YANKEE class SSN in 1984, the MIKE class in 1986, the SIERRA in 1986 and the ACULA in 1988.

In a decade that amounts to six totally new classes of submarines, and realize that when the SSN21 becomes operational in 1995, it will have been 20 years since the SSN688 had its IOC.

Collectively, we could say if we account for the new Soviet submarines, that the Soviets have the fastest submarines, they have the submarines with the highest power density because of liquid metal cooled reactors, they have the deepest diving submarines, they have the submarines that are hardest to sink, they have the largest SSBN and SSGN and the only non-magnetic submarine made out of titanium.

We have the safest and quietest submarines and the best submarines overall. If my son were a submariner, I would want him to be on a U.S. submarine. They have a lot of older, less capable submarines. All of the best ones don't represent the force capability. So looking at our force compared to their force, looking at their missions contrasted with our missions, we have a superior submarine force not only in terms of a submarine but in terms of the excellent people that command and operate those submarines.

What are the Soviets up to with all this capability in submarines? I believe that the Soviets think of their submarines or some of them as being capital ships, and certainly the capability that they are emerging with should give us cause for concern, and it certainly has given this committee cause for concern.

With those submarines, as pointed out by your advisory panel, the Soviets pose a serious threat to the sea lanes of communication. They also pose a severe threat to carrier battle groups and surface action groups. They pose a potential threat to our SSBNs. They pose potentially a submarine-launched cruise missile threat to CONUS.

They might be tempted to launch difficult-to-detect strategic submarine-launched cruise missiles against CONUS as one way to counter an effective strategic defense against ballistic missiles.

The Soviet SSBNs seem to be destined to operate in bastions where we have difficulty finding and holding those submarines at risk, and we can think of those SSBNs as being a survivable strategic reserve along with other strategic missiles of the Soviet Union.

What should we do to counter this capability? We should get on with the SSN21, as the first order of business. We should accelerate development of advanced submarine technology that is applicable to the new SSN following the SSN21. The Congress has taken the steps to do that. Also that technology can be used to upgrade existing submarines, the 688 and possibly later the SSN21. In the future, we should shorten the time between introduction of new classes of U.S. submarines.

Twenty years is too long when you compare that with what the Soviets are doing. Importantly, we should pursue a broad ASW initiative, and that should focus on the near-term, the mid-term and the far-term, and the piece of that that DARPA is most concerned with is the far-term and the exploitation of advanced technology to improve our ASW capability.

In that thrust and that initiative, we need to find new ways for surveillance of Soviet submarines for localization and attack of those submarines. We need to take a combined arms approach, to use a term usually used in connection with battlefield Army systems—we need to avoid platform constraints and we need to be prepared to change the balance across various Naval platforms.

I think the advisory panel has talked about architecture, and what I believe they mean by that is the distribution of capability that can be used in a team work approach by all of the platforms that are involved in antisubmarine warfare.

Finally, we need to pursue prototypes as Dr. Perry has suggested to the committee. DARPA has established an ASW plan. We believe that addresses the future threat, projects the worst possible case, but we think it is a plausible threat.

We put that plan together, drawing on the country's greatest talents in terms of technical aspects of ASW. It is a well-thought-out and broadly based plan.

Let me outline what that program would do. First, we would pursue very large two-dimensional Passive acoustical arrays that could perhaps exceed the capability we now have in the SOSUS system.

In the future, we will need very large two-dimensional arrays. We would pursue advanced active acoustics complementary to the work of the Navy. We would exploit computational technologies to develop and demonstrate automated acoustic detection and classification.

We are not going to be able to continue using people any longer just looking at displays to do our submarine detection and classification. We have to automate that if we are going to remain in the undersea surveillance and ASW business.

We need to explore acoustics and exploit unmanned undersea vehicles in order to get a better distribution of capability under water. We would use unmanned vehicles to complement and operate in affiliation and support of the manned submarines.

We would pursue automatic means for contact prosecution. That is a euphemism for localizing and killing the target submarines. Our approach is to build on the technologies that DARPA and others have developed for other applications.

We would draw on manufacturing technologies in order to make these large two-dimensional acoustic arrays affordable in terms of the hydrophones and the electronics for processing hydrophone data and stimulate broad participation and innovation, bring in new blood from industry, universities and from national laboratories of the DOE.

We would operate independent of but in coordination with the Navy. We would apply streamlined management methods which allow efficiency overall and allow DARPA, with its small staff, to manage a major program.

We have done that in the past and we think we know how to do that.

Let me turn to the advanced submarine technology program. We think the program is well-established. We have submitted three reports to Congress and we have briefed the staff. We have kept the staff informed about our progress, our problems and our plans.

The objectives of the program are to continue to improve the stealth technologies for our submarines, to improve the survivability of our submarines, to dramatically increase the payload capability, to decrease the cost of our submarines in the future and to reduce manpower requirements by application of automation technology in those submarines.

We think it is also important to improve sensors and weapons and to expand the mission capability of attack submarines significantly beyond ASW into new mission areas, but that is not part of this program that we are working with today.

Looking at some of the near-term results we expect from the program, we expect in the near-term that we will have established the best hydrodynamic computational capability that the country has within the Navy establishment at David Taylor. The capabilities will be far beyond what exist in the aerospace industry, and is adaptable to the design of submarines. There have been problems in the past with submarine design which you can attribute to the infidelity of the calculational capability.

The electro-optical periscope, which offers flexibility in terms of the space and the arrangement—it avoids full penetration and it also provides more information. It can be multi-spectral in nature and have infrared capability and visual capability.

That is a near-term capability. We are pursuing composite materials that offer a five-to-one improvement in strength to weight capability advantage for application to components in the near-term and in the long-term possible application to the hull of submarines.

[Deleted] demonstrate a [deleted] applied to a 688 submarine.

In terms of long-term results, we are going to maintain the stealth advantage of U.S. submarines. That is the most important parameter and attribute of a submarine, and we believe we can maintain that by an order of magnitude or greater compared to Soviet capability.

We are going to deliver a six-fold increase in the payload for the same submarine size. Our technology, we believe, will lead to that. We are going to make it possible to use sort of [deleted]. This gives a major increase in the detection or detectability performance of the submarine sonar system.

In conclusion, I personally thank the committee for its vision and for its continuing concern about this important area, and I believe that DARPA brings a unique capability and track record that can be applied to this serious problem.

Thank you, Mr. Chairman.

PREPARED STATEMENT OF ROBERT A. MOORE

INTRODUCTION

Mr. Chairman and members of the subcommittees, I appreciate the opportunity to appear before you today. I will address two extremely important undersea programs in DARPA: our Anti-Submarine Warfare (ASW) Technology Program and the Advanced Submarine Technology Program.

Our capability to detect and destroy submarines – to exert undersea control – is the subject of the ASW program, whose scope extends over sensor, weapon, and platform technologies and systems. The objective of our Advanced Submarine Technology Program is the development of innovative technologies that will enable future submarines to be more survivable and stealthy with greater firepower at reduced cost than is currently possible. It is important to recognize that these two objectives are very different. Just as submarines have important missions that go beyond ASW, ASW requires systems other than submarines.

The current budgets for these programs in FY89 are $79 million for ASW and $95 million for the Advanced Submarine Technology Program. Because of budget constraints, we were unable to include the Advanced Submarine Technology Program in the biennial FY90/91 budget, as you know.

The main themes of the report prepared by the House Armed Services Committee's Advisory Panel provide a reasonable framework for my remarks; namely, the Panel reported:

- Maintaining the superiority and effectiveness of our submarines is vitally important. .

- Detection and defeat of quiet enemy submarines is "a matter meriting the highest priority attention."

- The ASW problem contains formidable technical challenges. Near-term improvements to current systems are needed, but to meet the predicted submarine threat, advanced technologies and new system concepts are required.

—   The Navy has the ultimate responsibility for submarines and ASW. They must field the capability to meet today's commitments and future threats. DARPA must also contribute by focusing on the high–payoff, innovative technologies that are key to meeting future system needs.

I agree with these findings of the Panel and believe that our undersea technology programs are well attuned to them.

Anti-Submarine Warfare Technology Initiative

As the Panel reported, the progress of the Soviet Union in fielding advanced technology submarines poses a severe challenge to U.S. antisubmarine warfare capability. The speed, toughness, and diving ability of the Soviet Union's advanced submarines are properly of concern, but the biggest and most threatening change has been their increasing quietness. Because our present overall ASW capability relies heavily on passive acoustic detection, reduction in the acoustic signal strength of submarines translates directly into a reduction in ASW capability.

The predicted erosion of our ASW capability is a matter of strategic concern. The United States is a maritime nation; our national security requires open sea lines of communication to our allies and to our trading partners. In addition, the sea offers a medium for attack upon the United States itself, through cruise missiles launched from off our coasts or ballistic missiles launched from waters near the Soviet Union. Our national strategy requires that we be able to prevent hostile submarines from attacking us directly, threatening our oversea commercial links, or restricting our ability to project power with our carrier battle groups.

The problem we face stems from a change in the technology of our potential adversary, not simply an increase in numbers or an improvement in tactics. The solution to this problem does not lie in only short-term improvements to our existing systems, but in long-term innovations in technology and in the development of new concepts for the employment of this technology.

The Navy has the responsibility for responding to this challenge. The Navy must field the systems, develop the operational concepts, and maintain training and readiness to contain the threat. In short, the Navy must accommodate all time frames: current operations, near-term evolutionary improvements, and adaptation of revolutionary advances in the longer term.

But DARPA has a major contribution to make as well. While maintaining good coordination and communication with the Navy, we are conducting an independent program, building on our strengths to explore advanced technology approaches that may entail more risk and may take longer to develop into deployable systems, but offer in return the chance for greater ultimate payoff.

With the aid of the best talent in the U.S., we have formulated an Initiative to respond to this challenge. I will outline the specific projects, but first let me indicate the considerations that went into our decisions as to how DARPA could best contribute.

First, there are no easy answers — no "silver bullets" in sight that will solve the problem quickly. An integrated approach is necessary for long-term success. The Navy needs systems that provide better detection, better inter-platform communications for coordinated operations, better localization and tracking, and greater lethality in attack. DARPA will pursue technological advances in all of these areas, but will focus its attention on technologies that can be integrated in an overall systems context to achieve high-payoff results. We recognize that our primary role is to develop technology, but we will take this technology to the point that practicality and the payoff of systems based on the technology are clear.

Building on our current program, we will establish the limits of performance that can be obtained, especially from new detection concepts that are inherently resistant to Soviet submarine quieting. Our research efforts will be thorough and we will not waste time and money attempting to develop systems that are not effective and affordable.

Furthermore, we will work closely with the Navy, so that we can better understand their operational needs, and so that they are aware of the implications of our work and the opportunities it offers.

We have presented the elements of our ASW Technology Initiative to the Armed Services Committee in a report submitted in February, so I will only briefly summarize the six major areas on which we are focusing our attention.

Passive Acoustic Arrays: Even though continued Soviet submarine quieting would severely reduce the performance of our fixed detection systems, we think that large, two-dimensional arrays may "buy back" a portion of the performance that we stand to lose. Three competing industrial design teams are now working to reduce this notion to a practical design. We must determine the ocean's limits on performance and while we have not established those limits yet, we expect to demonstrate useful detection ranges even against the "very" quiet threat. We intend to explore these limits by fielding very large experimental arrays and developing advanced methods and algorithms for processing the signals. We will also investigate new engineering methods and manufacturing techniques to lower costs — a requirement for systems having thousands of elements.

Advanced Active Sonar: In principle, active sonar is not as vulnerable to submarine quieting as passive systems since it provides its own acoustic signals to "bounce" off the target. Important programs are ongoing in the Navy, but a number of signal processing concepts require additional examination. This year we have developed a new theory pertaining to ocean noise, which, if confirmed by experiment, will allow major system performance improvements. Cost effective systems will require development of technology for very powerful low-frequency active sources. We intend to work on both of these areas.

Automated Acoustic Detection: If we are to detect quiet submarines in all of the environments of interest, advanced signal processing approaches must be developed. This will require processing enormous amounts of data with completely automated processing systems. Some detection concepts will require autonomous systems that can perform the complete detection and classification process without human intervention. Developing this capability will stretch our abilities in high-throughput computers, advanced signal processing methods, and artificial intelligence. We have efforts underway in these areas for acoustic arrays that adaptively process the environmental and shipping noises, "smart" processors with neural nets to automatically detect non-traditional signals, and sonar fusion and detection automation systems.

Non-Acoustic Remote Sensing: For all of the effort that we intend to put into improved acoustic sensing methods, we cannot neglect other signals. If we can discover how to detect submarines remotely – by space-based radar, for example – then we will have a surveillance capability with entirely new characteristics that could have a substantial impact on anti-submarine warfare. We do not know yet whether such a capability is possible, but we and others are doing the research to find out. We have been working in non-acoustics jointly with the U.K. and Norway as you know, and we are close to concluding MOU's for continued cooperative efforts. DARPA and Navy have also begun to test a new non–acoustic anti–submarine warfare system which has high potential.

Unmanned Undersea Vehicles (UUV's): Fully autonomous or remotely piloted vehicles may give us unprecedented capabilities to employ sensors and even weapons in areas where unaided submarines would face high risk. Limitations in computer capability and artificial intelligence for autonomous operations, and in power systems have prevented their use until now. Development of the necessary technology has been underway in other parts of the DARPA program and we expect to complete construction of our first prototype UUV later this year. There is a good deal of work still to be done, but we are confident that UUV's will provide a valuable asset to Navy's forces in the next decade, and we will push aggressively to develop them for specific, high–payoff applications.

Automated Contact Prosecution: Even with success in the advanced detection methods that the Navy and DARPA are working on, it seems likely that very quiet submarines may elude detection except for very short intervals. This means that we must localize and attack the target in much shorter times than is possible today. As a consequence, aerial delivery of the weapon system is almost certainly necessary and re–localization of the target at the end of the flight may have to be autonomous. We have identified concepts for doing this, and we will develop them to the point that we can demonstrate their performance. If we are successful here, then we will have the key enabling technology for a long–range weapon system that could prosecute targets detected by any of our advanced surveillance systems. This new capability would have substantial payoff in ASW, particularly in forward areas where our ASW platforms and systems have the most difficulty operating. We have started a program which, if successful, will lead to the development of torpedo warheads with improved lethality, an airborne delivery vehicle that will get them to the targets, and a localization concept deployed from a small aerial vehicle which offers substantial promise.

These six project areas are not new for us. We have been working in related areas, and these programs build on the science and technology that we have developed in high speed computing, acoustic and non-acoustic sensing, unmanned vehicles, and signal processing methods. These projects were selected with the aid and participation of the best talent available in the U.S. They offer the highest prospects for large payoff in ASW, and they are ready to be pushed hard. The ASW Technology Initiative, as presented in my report to you, describes a high-risk, high-payoff technology program which we intend to pursue at as fast a pace as available resources allow. The program will require the best efforts of talented people both in and outside DARPA, and it will require a sustained effort over an extended period of time with stable funding.

Advanced Submarine Technology Program

The Advanced Submarine Technology Program was established by the 100th Congress and it is exploring a broad range of innovative technologies to provide options for future submarines. The program is of critical importance due to the roles that submarines play in our national defense and the fact that the Soviet Union continues to demonstrate significant progress in submarine capabilities with a concomitant expansion of mission and roles. It is this Soviet progress that alarmed the HASC Advisory Panel, which recently issued their report on Submarine Technology and Anti-Submarine Warfare. The report summarized the situation with the statement:

"Though we saw no evidence as yet that the Soviet's ambitious R&D program has produced a submarine that is qualitatively superior to ours, their continuing, ambitious technology program may well produce such a submarine unless our own R&D efforts at least match theirs in scope and productivity."

It is important to note that the Advanced Submarine Technology Program at DARPA is addressing new technology that can be useful not only in ASW, but across the full spectrum of missions that a submarine may be called upon to execute. A primary example is stealth technology, which will allow our future submarines to operate in forward, contested areas with little fear of detection.

The budget for this program was established by Congress at $113 million for FY88 and $95 million for FY89. It is proceeding according to this original plan and making excellent progress. The program now has 195 contractors, 9 universities, 3 national laboratories and 5 Navy laboratories working. As I indicated before, funding of the program in FY90 and beyond is not included in the budget request.

Program Rationale

The Advanced Submarine Technology Program is addressing innovative technologies that will improve the key performance attributes at a reduced cost. The submarine's overriding attribute is its ability to remain undetected, i.e., stealthy, during selected portions of a mission. Following stealth, two equally important attributes are survivability in an engagement and effective payload delivery. To optimize the value of these attributes, the physical characteristics such as speed, operating depths, and acceleration must be analyzed using a systems approach and mission requirements.

Program Objectives

The objectives of the Advanced Submarine Technology Program remain as reported to Congress in our recently submitted Long Range Plan:

-   Identification and development of promising and revolutionary technology that can provide new and innovative design options for future U.S. submarines;

- Mobilization and focusing of industrial, university, and government research and development talent to significantly improve future submarine performance;

- Demonstration of advanced technologies and systems by rapid prototyping

Program Achievements

. In response to an April 1988 Broad Agency Announcement, we have received and evaluated almost six hundred proposals from industry, non–profit organizations, government laboratories and universities. The total value of all proposals was over $4 billion. Only the proposals that offered the highest payoff were selected for award. The message has gotten through to industry— we are looking for innovative technology and advanced concepts even if they conflict with the "conventional wisdom." An example is magneto-hydrodynamic or MHD propulsion which may propel our future submarines silently without the need for a propellor, reduction gears or drive train.

The program has established a forum for visionaries using workshops and symposia. In addition, the program has pursued university involvement to train new engineers and scientists who have an interest in submarine development. Furthermore, university involvement is occurring through a cooperative effort with the Office of Naval Research.

Technology Timing

It is particularly important that I comment on the implications of timing and long–term commitment toward the development of future submarines. A number of enabling technologies being explored in the program, such as superconductivity, vastly improved computer power, and advanced materials may allow us to make order–of–magnitude changes in how we build, operate, and maintain submarines in the future. However, much of what we are doing today will not be available for a decade or more. In other words, the submarine technology program is not in competition with the Navy's on-going building programs such as the improved SSN688 and the SSN21; but complementary, when viewed in the time-frame of the next twenty-five years. As I will discuss later in more detail, from a perspective of timing, it is important to establish a firm basis on which to transition technology to the Navy that is compatible with their resources and needs.

The DARPA/Navy Program Relationship

The Advanced Submarine Technology Program is consistent with our DARPA charter to pursue high-risk technology that promises high utility in military payoff and allows us to explore revolutionary technologies. Having the Advanced Submarine Technology Program at DARPA is consistent with the HASC Advisory Panel report recommendation that research on advanced submarine technology should continue in both the Navy and DARPA, and that the two should be independent in approach, but not in planning. We can be "constructively adventurous" and move out ahead of formal requirements to pursue the risky course usually shunned by others; in addition, the Program directly benefits from the synergy of other ongoing DARPA efforts in materials, superconductivity, unmanned undersea vehicles, laser communications, and advanced computer technology.

I am pleased with the degree of coordination taking place between DARPA and the Navy in this Program. The Navy's recent initiative discussed by VADM Cooper in his March 2 statement to the Seapower and Critical Materials Subcommittee to establish an integration effort for advanced submarine technology development through an advanced submarine R&D office, NAVSEA 92R, is an example of their cooperation. Adequate funding of this effort is necessary to efficiently transfer DARPA's technology developments to future systems. We intend to work closely with NAVSEA 92R to stimulate transition planning that will take advantage of what DARPA has done and smoothly fit into the

Navy's acquisition cycle.   Our mutual efforts are being close-coupled to prevent "re-inventing the wheel" and to maximize our finite resources.

One of the biggest challenges for the program will be demonstrating the emerging technologies to the necessary degree that they can successfully transition to Navy programs.   Projects like electric drive, submarine automation systems, composite hulls, and fuel-cell power systems are expensive to bring to fruition.   The Advanced Submarine Technology Program is a mechanism to sustain these high-payoff projects that will yield improved submarine forces for the 21st Century.   We appreciate the continued support of the Congress.

Conclusion

I hope I have conveyed to you the importance of both the ASW program and the Advanced Submarine Technology Program as separate but complementary elements of a DARPA Undersea Warfare Technology Program.   With the support of Congress, we have made substantial progress.   However, the challenges facing us in technology and resources are severe.   We must make a sustained commitment to a national program, executed by both DARPA and the Navy, if the challenge of the quiet submarine threat is to be met.

Mr. BENNETT. Thank you.

I think we will go through a few questions while everybody is here and then we will go into a briefing, because I don't have enough to exclude from the hearing everybody. After we have a chance to inquire, we will hang around a little bit and go into a closed briefing.

I must say the thing that concerns me about ASW is the announcement, which is true, that ASW is on the front burner of our concern of being with the kind of Navy we ought to have when that is accompanied by an apparent lack of dynamic leadership of somebody on the civilian side who is assigned this duty.

When I came up with whose the czar, the answer was, first of all, the CNO. Under the Goldwater-Nichols bill, there is supposed to be a civilian who is running this sort of thing. When I asked who this was, it ended up being the Secretary of the Navy. The personnel of these two spots are wonderful people, but with all the administrative responsibilities that they have in the Secretary of the Navy's office and all of the strictly military things that are involved in the CNO's office, I don't see somebody I would really like to see here performing today. Not running you down, but I don't see anybody who is really carrying this ball which I thought our statutes in the Goldwater-Nichols bill prescribed, and the civilian people who had that job are no longer in the Navy, I understand.

That disturbs me a bit. I would like you to address that. We can do it in closed session, but I thought I ought to raise it here because I don't see the dramatic language of our concern being matched by somebody like Rickover or Lehman or somebody who is running with the ball. I don't see one person who is really making a real uptight thing he is working on.

Admiral SMITH. Your point, Mr. Chairman, is an interesting one. I am sure you realize that in the procurement side, the acquisition side under Goldwater-Nichols, of course the Navy has designated the under secretary of the Navy as the acquisition executive, so he has the responsibility and executes it for those procurement decisions that relate to all of our acquisition problems.

Mr. BENNETT. Since it is our first order of business and the greatest worry of the Navy, shouldn't it be put in the hands of somebody who is just dedicated to this effort?

Admiral SMITH. We do have people dedicated to this effort.

Mr. BENNETT. Who is the top line civilian that is running it? I thought those places were open and the person had left.

Admiral SMITH. Mr. Chairman, the current fully dedicated R&D individual we have for ASW is Admiral Wolkensdorfer.

Mr. BENNETT. He is wearing the mufti.

Admiral SMITH. But he is fully dedicated.

Mr. BENNETT. But he is not, according to the Goldwater-Nichols bill, a civilian who is running this. He is not in the same position of independence from the hierarchy above him.

Admiral SMITH. The individual coming closest to representing that would be Mr. Pyatt at the platform level. He is thoroughly familiar with the acquisition programs that involve our ASW systems. That is the SSN-21, it is the SSN-688, it is the advanced improved sonar system for our surface ASW platforms as well as the Arleigh Burke destroyer class, and it is the update to the P-3 aircraft, which is our primary ASW air weapon. So in the civilian side, we focus the acquisition in Mr. Pyatt under the under secretary of the Navy.

Now, across the spectrum it is the chief of Naval Operations' responsibility as delegated by the Secretary of the Navy for the requirements that come from our warfighting base. The master plan of ASW determines what that requirement process produces, and that requirement process then inputs to the acquisition process those requirements that we based on experience and on the advice of the expertise in the R&D community are the next generation systems that we could develop that are technologically feasible, that are effective if we could get them to the operational stage.

We push them into the system and then the acquisition process produces that particular piece of hardware. There is no program manager concept.

Mr. BENNETT. I kind of think there should be. I think there should be somebody who is dedicated to this job, since it is such an important job. I don't want to trespass any further on people's time. We have a list of other people.

One other thing I want to say because in the last hearing some people had the impression that I thought there should be an amendment with regard to the testing of the submarine. The law that was passed in 1983, which I was not the author of, although I favored it, said that everybody ought to be tested before you go into high production and it left low production to the definition of Congress.

Since we are just producing one submarine a year, that isn't exactly speedy production, so I think Congress can pass on it as it goes along. I don't think the law needs fine-tuning.

There is nothing dramatically different about the hull of the 21. Therefore, we could go ahead with the production of the hull without testing that particular hull or we can simulate it. The internal workings of it should be tested as promptly as possible. It seems to me that it is not likely we need fine-tuning even in the law we now have, because it says low production, and one a year doesn't strike me as a very high production rate.

Is that what you feel about it?

Admiral SMITH. I think it would be the Navy's view that the hull, although it is a unique shape, is identical in construction to earlier hulls. It will use higher strength steel to give it an additional operational envelope, although the shape will be different. The test of that hull is accomplished during the first series of operational trials of the ship, which only take 3 or 4 days.

Mr. BENNETT. The way in which the rules of the House require— Mr. Spence, would you like to address questions?

Mr. SPENCE. Mr. Chairman, thank you.

I was just trying to think back on the report issued by the so-called panel of experts and reconcile that with some of the things in the statement here this afternoon, and from the standpoint of our priorities. As I remember, the report indicated that maybe our priorities are misplaced without enough emphasis being put on ASW.

Here on page 5, I read in your report, that ASW is our number-one warfighting priority. I am just trying to reconcile that.

How can they find that there is not enough in the way of emphasis being placed on ASW, when you say that is our number-one priority?

Admiral Smith, that was in your statement. That was the main thing I was trying to get reconciled.

Admiral SMITH. I would like to ask Admiral Jim Fitzgerald. He is from the director of Naval Warfare's Office and they are the requirements determination and the priority-setting organization. I would like to ask him to address that question for you.

Admiral FITZGERALD. I am going to have to ask you to rephrase it. I am not sure I understand exactly what you are asking me.

Mr. SPENCE. The panel reported back to us that ASW didn't have the top priority place within the Navy it should have.

Admiral FITZGERALD. Although we said it did, is that the issue?

Mr. SPENCE. You said that is our number-one warfighting priority. I don't know whether they overlooked something or what. Do we know more than they know?

Admiral FITZGERALD. I think the issue of ASW as the number-one warfighting priority goes back to the early 1980s when the Soviets began the particular quieting evolution that we detected. We initiated a series of efforts which began with ASW master strategy group that was composed of Navy people, industry people, people from all the Navy laboratories and civilians, and developed an ASW master strategy.

From that came an R&D development strategy. Out of that came 15 different operational requirements that needed to improve our ASW capability.

We have iterated through that, formed the ASW assessment team to study that. It is an ecumenical body including representatives from the DCNOs, DARPA, CNA and the different laboratories, and we have developed on to the third edition of the master plan, which includes development strategy, future concept of operations, how we see ASW out to the year 2010, top level warfighting requirements, what we have to satisfy to carry out the national strategy.

The issue was that the amount of resources that were being dedicated to it were—perhaps the panel thought they should be more, and we thought commensurate with all the other priorities we had, that they were somewhat less than what they would like.

Mr. SPENCE. In other words, the panel said, start over, you are using the wrong approach. Even though we have spent a lot of money in that particular area, we are going down the wrong track and need to start over.

Admiral FITZGERALD. ASW is evolutionary. There are no silver bullets. ASW is an immensely complex problem which depends upon the environment. You can't say we will buy a submarine or airplane and that will solve the problem. It is like a jigsaw puzzle; it takes a hundred pieces to make the whole picture, and you may have 45. On a good day you have 30. You have to understand oceanography, the physics of sound and water, and that sort of thing. So it is complex.

Mr. BENNETT. Mr. Foglietta.

Mr. FOGLIETTA. Thank you, Mr. Chairman.

If the ASW, as you say, has such a high priority, why isn't there money in the 1990 budget to continue the DARPA study?

Admiral SMITH. Would you repeat that?

Mr. FOGLIETTA. I said, if the ASW program does have such a high priority, how come there is no money in the 1990 budget to continue the DARPA study of ASW, unless I am mistaken and there is money there.

Admiral SMITH. The Navy budget does not contain any of the funding for DARPA. The DARPA budget itself completely contains the DARPA funding, and I would think that question would probably best be handled by Mr. Moore.

Mr. MOORE. Congressman, I am not absolutely certain which study you are specifically referring to. We certainly have an ASW budget and over the past 2 years, we have increased the funding for ASW.

Mr. FOGLIETTA. Let me rephrase the question. Is there money, are there sufficient funds in the 1990 budget for the ASW program within DARPA?

Mr. MOORE. I support the President's budget, but from my point of view, we have defined an ASW high technology initiative which we think is justified and we defined in the report we submitted the additional funding required. I think that is a good start.

It is a very big problem. There is a lot of technology that is applicable that needs to be explored, and additional money would be required in our budget in order to be able to prosecute this program.

Mr. FOGLIETTA. I have got some information here from the staff, but I think that information combined with your answer satisfies my question.

My second question to the Admiral, what technological capabilities do the Trident submarines have to counter the Soviet attack submarines you talked about earlier?

Admiral SMITH. I would say the fundamental property that the Trident submarine has is it is very, very, very quiet. It is quiet to the point that the stealth is enhanced by its ability to operate at a reasonable depth—I mean not near the surface—and at reasonably slow speeds.

Within the Trident design originally was an iterative improvement to the propulsion plant that made that, at the time of its construction, the quietest propulsion plant in the submarine business.

Our fundamental strength remains the acoustic spectrum. You can't hear them and you can't find them.

Mr. FOGLIETTA. Thank you.

Mr. BENNETT. Mr. Kyl.

Mr. KYL. Admiral Smith, on page 8 of the panel report it reads: "After months of struggling with the issue raised by the DSB, the Navy leadership has proposed a plan for reorganization. Though it would create a czar to manage the Navy's ASW R&D efforts, the position will be only at the two-star level and thus still inferior to the dominant platform sponsors. We cannot tell at this point what effect, if any, this reorganization will have, but it seems at first glance to be embroidering the organization chart rather than attacking the root problem of why the Navy's most serious challenge is of a lower priority than business-as-usual issues such as force structure."

I would like you to comment on that.

Admiral SMITH. I think we would not agree with the impression that the two-star admiral is below the platform sponsors. He is the, probably the senior Naval officer in the Pentagon dedicated to ASW basic exploratory and advanced development.

He is not working across all the platforms. He is not working integration, which is being done in the director of Naval Warfare's office. He is not making platform trade-offs. He is worrying the ASW problem from an R&D standpoint, and the future of that ASW R&D initiative.

In that context, because he works for the Assistant Secretary of the Navy for Research and Engineering Systems, that is where the money flows from for all of this R&D effort. So although he is a two-star, he has his hand on the value that controls the funding dollars for that R&D and that makes him in R&D more powerful than the platform sponsors because he is controlling the flow of those dollars.

Mr. KYL. Seeing that this program develops properly in line with the seriousness of the challenge, I think it is very important to keep that concern of the panel in mind and hope to have the plan operate as you have said.

In your testimony there were two other things I wanted to ask you about. You indicated on page 1, reports, fleet performance and congressional hearings show also the Navy has recognized fully over the last three decades the importance of maintaining our ASW advantage.

The panel report suggested that while technical problems and Soviet advancements are the chief reasons why we have gotten into the situation we are in, nevertheless there are important things that the Navy should be doing that suggests that they didn't quite agree with this conclusion in your testimony.

Do you want to take a minute to respond to that point?

Admiral SMITH. Your comment is that we are not——

Mr. KYL. With all due respect, and I mean this with the utmost of respect, your written testimony puts a little bit nicer light on it than the panel did.

We have recognized these problems over the years. The next statement says, the Soviet submarine force is large, quiet and formidable, but the results of a recent ASW technology net assessment reveal positive indications that our programs are working.

Later you refer to the fact that by the year 2000, the assessment concluded the U.S. has a qualitative superiority in submarine platforms. the panel says that unless certain things are done, that isn't going to be the case.

Admiral SMITH. We would have to say we don't agree with the panel. I think you should realize that we are looking across the entire spectrum of ASW. In the year 2000 we are pitting the 688 class submarine against—we have those in relatively large numbers. There will be more than 35 in the fleet at that time—against a very few modern generation Soviet submarines. The bulk of their submarines in the year 2000—70 percent of them are of the older type submarines, so we do believe that we have the advantage not just because of that number change, but because of the qualitative capability of the submarines.

Mr. KYL. Let me be specific and refer to the exact language of the report.

On pages 6 and 7: "The introduction of quiet Soviet submarines could well lead to a fundamental change for us in naval warfare. Unless ways to compensate for the coming degradation in passive sonar can be found:

"Our attack submarines are likely to lose much of their effectiveness in what has been one of their major missions—operations against Soviet submarines;

"Because they will be so much harder to defend against, Soviet attack submarines are likely to become far more effective in operations against our surface ships, both naval and merchant; and

"Deprived of cuing, both our surface warships and our aircraft could have great difficulty in open-ocean ASW.

"As noted above, these developments—which seem likely in the absence of adequate replacements for our traditional passive sonar systems—threaten the survivability of our carrier task forces, our ability to deploy and support our military forces overseas, and the security of our coasts against submarine-launched missiles."

These are extraordinarily ominous predictions, particularly coming from this panel of experts and recognizing that they didn't necessarily say this would come to pass; they said it would be likely if we can't find a way to cope with these quiet Soviet submarines.

Do you agree with that assessment?

Admiral SMITH. Yes, but I would also have to state in support of our statement that we are attacking and developing solutions for the problems that are inherent or underlying those predictions.

Mr. COLEMAN. Thank you, Mr. Chairman. I will have some more questions later.

Mr. BENNETT. Mr. Pickett.

Mr. Bateman.

Mr. BATEMAN. Thank you, Mr. Chairman.

I have a couple of things that I hope, I will have time to inquire about. The first of them, there are some who apparently are contending for budgetary or for whatever reasons, that maybe the things that should be given up are the last two of the 688 class submarines.

It occurs to me that is probably a very bad idea as you look at the quality of the latest model, so to speak, of 688 against the much more inadequate Soviet submarine threat it would be deployed against over the next probably 20 years or more, and the cost of those two additional and final 688s relative to anything else that we would be able to obtain and deploy against the same quality of threat within anything like the same time frame.

I would be happy to have your observations on that and then, Mr. Chairman, there is a secondary that returns to a subject that you broached on the so-called prototype testing and where we stand vis-a-vis the development of the SSN-21 and somebody's interpretation of the prototype testing bill.

I hope, Mr. Chairman, your interpretation is good, and I don't think there is a better legal mind and interpreter around than you are, but it frightens me that if the people who are going to make the official interpretation don't have your interpretation, that we may put ourselves at considerable risk if we don't do something to make it clear that your interpretation is indeed the interpretation that the Congress intended and wants acted upon.

I would like to hear any observations from the panel that you have with respect to this testing issue in the context of the SSN–21.

Admiral SMITH. Mr. Bateman, I think in the case of your explanation in defense of the 688 class submarines, we would certainly support that and it is the Navy's intention to do everything that it can to retain the submarines that are currently in the budget.

With regard to the testing, I think you would find on even just a casual examination that the SSN–21 is probably one of the most thoroughly tested, thoroughly worked out submarines that the Navy has built of any kind of ship.

We have a quarter steel model test going on to test the hull form and the propulsion design studies. We have done expensive work on individual compatible systems by testing them on other submarines with testing the increased strength hull.

Mr. BATEMAN. Admiral, I really don't think that gets you responsive to the area where you may indeed have a concern. Certainly, I know from years on the subcommittee, how much testing and study and work has gone into making the SSN–21 the submarine that I have confidence that it is going to be. It has been tested in every way, but we are dealing with a statute passed by the Congress saying, you must operationally test a system before you can contract for even further production.

I think it is going to be a disaster for our security if we can't produce a second SSN–21 until we have built the first one and operationally tested the completed constructed operating submarine for a year. That is the legislative context in which I think you may have to operate, unless we get clarification that whoever is going to make the final and official interpretation of the statute is going to do it in the rational manner that our Chairman or that you or I maybe would.

There is apparently a body of official legal opinion that is frightening in the context of what happens in the SSN–21 program.

Admiral SMITH. I would say that low-rate initial production, the word "rate" has some implication other than one data point, so that buying one of a product does not give you any kind of rate production. Certainly, in looking at the current budget, there are no submarines of the SSN–21 class in 1990, and two in 1991, so that is one submarine a year. If that is not a low rate, then I am not sure how to define it.

Mr. BENNETT. That is all the law says. It just says a low rate. Any law that is on the books can be changed by the next Congress, so whenever we get to a place where people can say we are having a high rate, we cannot build that shape that year.

There is nothing to prohibit you from doing what Congress authorizes you to do; it just gives a direction it will be a low rate. I think it is a low rate.

Mr. Pyatt is going to give you a legal opinion as to whether he feels any fine-tuning is required, but there is nothing requiring you to produce a SSN–21, get it in shape, and then test it before you produce another one. Nothing like that is in the law.

Furthermore, if we really did do something as dumb as that, we would be throwing away a decade of opportunity to be ahead of our potential opponents; in other words, a whole decade of technology would be lost. I didn't want to take that time away from Mr. Bateman, so I will put it back on.

Mr. BATEMAN. Mr. Chairman, if I may reclaim 30 seconds of time to say I compliment you on your statement. I am glad it is in the record. I hope that those who are going to be making any legal interpretations will take time to read this record. If they come down to any other conclusion, that indeed this committee and this Congress will straighten them out forthwith, because it would be unthinkable not to go forward with the SSN-21 as promptly as we can put that submarine in the water.

Mr. BENNETT. Thank you very much.

The next person to inquire will be Mr. Darden.

Mr. DARDEN. Thank you very much, Mr. Chairman.

I thank the panel for being here and appearing before us today. I am not sure toward whom this question should be directed, but with all the expertise in front of us, I am sure there is someone who will have the answers to this.

Gentlemen, last year, our committee report on naval underseas warfare, R&D noted the dire consequences of the effectiveness of passive sonar due to the quieting of Soviet submarines. It is for this reason the committee recommended last year an authorization of about $5 billion to investigate the potential of the SB-22 in concert with active sensors that could be deployed, retrieved and then redeployed.

The $5 billion was appropriated in the fiscal year 1989 appropriations bill. In spite of our obvious interest in the potential of the SB-22, it is understood that the Navy now plans to reprogram those funds for something other than sea-based air ASW.

As I recall, the funds, Mr. Chairman, were in the House bill. They were not approved by the Congress in authorization, but they were in fact appropriated.

My question to the individual on the panel who would be familiar with this area is: In light of our past findings and our report on ASW, would you explain to us the Navy's decision to divert these funds from their intended purpose? Since reprogramming approval is not a foregone conclusion, could you tell us on the committee how the Navy would use this $5 billion to develop a new approach to sea-based air ASW capability?

Admiral SMITH. Mr. Darden, we will have to take that question for the record.

Mr. DARDEN. Very good. I realize it is a narrow question, but it is something that we have been interested in on the R&D committee in particular.

That is all I have, Mr. Chairman. Thank you very much.

Mr. BENNETT. Mr. McCurdy.

Mr. McCURDY. Thank you, Mr. Chairman.

Following the chairman's lead on his original question, trying to find out who is in charge of ASW, I said somewhat in jest that it is hard enough to figure out the Navy's organization when you are wearing your winter uniforms, your rank, but I was interested because Admiral Evans was by here talking about SEANAV-92.

I was trying to figure out what the code word was on the numbers. Maybe somebody would care to explain that to us. It is obvious that it is—that there is no clear organizational line that says this person is in charge of ASW. I think the gist of this hearing, the gist of the brief R&D hearings, the gist of the policy panel and all the other subcommittees that have been meeting on this is we consider this to be a very serious issue, that the outside panel's report, the fact that most panels come out with the minimum approach, that is the lowest common denominator.

They try to come to the one point where they can all agree. When they come with a report that states their concern about it, it is obvious that this committee and others in the outside community consider this to be a grave national concern.

If that is the case, and I think the Navy has come along and agrees with that, but we have had somewhat of an experience of reluctance in dealing with the Navy, from the Navy's standpoint.

The gentleman from Arizona, Mr. Stump, and I worked in creating the joint project, a look at space-based ASW and other types of ASW and efforts the Soviet might be involved in. Now, everyone admits that was a worthwhile endeavor and that it should be continued. When we started it, everyone said, no, we shouldn't have any outside approach.

We then went and we got involved with DARPA and we directed them to do this work, which is ongoing. Yet, quite frankly, it appears to us, at least to this member who has been following this for some time now, that DARPA may have too much on its platter here, that DARPA perhaps—and I know the NAVSEA-92 has its bag open waiting for DARPA to drop those funds into that bag.

But at the same time, the question is, who is going to set the priorities? Who sets the direction? How are you going to agree on this?

Let me give you some examples. You have got the problem in the TSUNAMI Project ongoing over here in one agency. That is not a classified term, by the way. You have DARPA. You have two or three organizational lines over in the Navy, yet there doesn't appear to be much coordination.

What we are telling you is we think there has to be greater coordination. Now, DARPA, for instance, in—Admiral Evans and I talked about this. I note the Soviet have been looking at this. The questions of composite holes, the question of structural acoustics, the construction of reinforced, plastic reinforced materials, the whole materials issue.

OK, now, who should be doing that research? Should DARPA be taking that and saying this is our, one of the mainstays, this is a critical program, and we ought to be out there on the cutting edge, because it is a costly endeavor, very expensive, takes a lot of research.

But you have got expertise in other materials, et cetera. Should DARPA be taking that and concentrating on that? Should our Navy in turn take all the $6.2 and start picking up the slack on superconductivity and other types of riskier research out in the future and other technical challenges?

Have you all reached an agreement on that? Who makes those decisions between you all? Please don't tell me it is Mr. Paisley, or Pyatt.

Mr. MOORE. There is a great deal of interaction now between DARPA and the Navy in all areas of the technology program, including the materials area that you mentioned. As you know, as directed by Congress, there is an advisory panel to DARPA on its advanced submarine technology program chaired by OP 98, which is the Director of RDT&E in the Navy, and with the Office of Naval Technology represented by its technical director and with the Director of Submarine Warfare being represented on the committee and other Navy representatives, there is a great deal of coordination and cooperation.

As you know, we have one of the finest naval submarine commanders, Captain Palaez, who is in the audience, who is our program manager. So I think in submarine technology, we are certainly, well coordinated, we are communicating. I think that we are beginning to develop a transition plan.

This program, from my point of view, as one who is in his third tour at DARPA and worked in various capacities at DARPA is an unusual one. It is somewhat extraordinary. It has been difficult, but I think we have gotten on top of it.

Clearly, there are parts of this program that belong in the Navy. When we look at the materials area, the Navy started working with composite materials before I heard about an advanced submarine technology program coming to DARPA. We have accelerated that.

We have brought in new blood. We brought in capabilities that you find in the aircraft industry, and we are trying to exploit that. As soon as we can, we are going to start transitioning part of that back into the Navy.

There are parts of it that we think are more straightforward, less risky that will pay off. Some of the components inside the submarine machinery bases and things like that should transition back into the Navy and I think it will.

Some of the more risky areas, the thick composite hole sections and so forth, we ought to continue with that for a few years and resolve the issues.

Mr. McCURDY. If I could—Mr. Chairman, with the committee's indulgence, I want to follow up on a point. I am not on DARPA's case. You will not find a more vocal supporter of DARPA than I am.

I think DARPA is doing a great job. What I am alleging is that DARPA is perhaps trying to do too much, because not enough is being done by the Navy. Let me give you an example. The Navy has about 14 to $15 million in 6.2 monies. How do you reconcile that with the submarine technology program in the DARPA effort? I mean, Navy has not—you can say you got a whole new division and new ops and new 92s and all these other folks out there, but the point is that the budget has not reflected this as a priority.

What we have been saying is that there is a priority, that this is a national priority. Now, we have not reconciled it; granted, my friend from Virginia talks about the 688, but he also talks about the carriers and aircraft and all the other. He has not met a ship he doesn't like yet.

They are all necessary, but that is not the question of priority. Who is going to decide? Either you are going to tell us that this is not as high a national priority as we think it is or you are going to change your budget.

You can't have it both ways. You cannot come before this sub-committee and tell us, "You are right. The Soviet have made all this improvement. You have got to get a handle on it," and then not put any money into it. Who is going to make that decision?

Admiral SMITH. The point is well taken. The CEO and the Secretary of the Navy are going to make that decision. They made that decision in the 1990 budget. When they made that decision in February of 1988, the Defense Science Board has just reported out with a report that said the amount of money being devoted to exploratory R&D, the 6.2 R&D in the Navy for ASW was deemed by them to be adequate, but was not properly focused.

Now, where they had——

Mr. McCURDY. Can you tell me how much money you are talking about? That decision was made. Give me a dollar. Can you just tell me a total package of how much money you are talking about?

Admiral SMITH. It is probably in the neighborhood of $20 million.

Mr. McCURDY. Twenty million dollars?

Admiral SMITH. Right.

Mr. McCURDY. You are telling me that this is a national priority, and we are going to put $20 million into it?

Admiral SMITH. I am saying this is the exploratory R&D piece, and there is another 4.5 or 6 billion going into the rest of ASW across the spectrum of the Navy, which is fielding systems, doing the advanced engineering development and bringing systems to fruition, is modifying and improving systems based on the fleet's own experience and support, and is then generating in the basic technology area other new ideas which are below that.

We are talking about one piece of a very, very large, complex problem. That is the piece that Wolkensdorfer is going to take over. That is the piece Admiral Evans is working partly, not totally. He is working that plus other initiatives.

Mr. McCURDY. Can I ask you another loaded question? You have got two Rear Admirals, Evans and Wolkensdorfer. Between them, what is the combined budget between those two Admirals?

Admiral SMITH. Let me defer that.

Mr. McCURDY. Please.

Admiral EVANS. Yes, sir. Congressman, nice to see you again. We had a good chat in your office. I appreciate the opportunity to talk to you and the distinguished members of this committee on this program.

I think in these discussions, we at times get a bit askew as we forget there are two separate initiatives at DARPA that are funded separately. One is the advanced submarine technology program, which Mr. Moore has referred to, which relates to the development of hull, mechanical and electrical technology for future classes of submarines.

The other is the anti-submarine warfare technology initiative, which is separately funded at a much lower level. The advanced submarine technology program is funded in the fiscal year 1989 budget at a level of $95 million. I am not sure what the number is for ASW. It has been much lower than that.

The myth that DARPA has not been involved in submarine technology before is simply that. DARPA has been involved in submarine technology programs in the past, on the leading edge with technology for high-risk types of things that have submarine application. The injection of the $100 million in fiscal year 1988 and 95 million in fiscal year 1989 by the Congress into the submarine technology program for the reasons accurately documented in various publications and records, including the legislation itself, are well-known.

So, I am responsible for working directly with the DARPA submarine technology program director to determine the opportunities for transition of those technologies that DARPA is working on, in the submarine technology program for further application by the Navy.

Another thing that is at times difficult to state clearly is the budget total. The budget lines that I am specifically responsible for in fiscal year 1990 include the $29 million in 6.3 development monies in a new line which is referred to as the Advanced Submarine Systems Development Line.

The focus of that line is, in fact, hull, mechanical and electrical initiatives, and was specifically designed to instill a new vigor, if you will, into the technology transition process, not only for hull, mechanical and electrical projects that will emerge from the DARPA program, but also those that will emerge from the Navy's 6.2 hull, mechanical and electrical technology block.

That line goes to a total of $54 million in the fiscal year 1991 budget, and then rises through the current 5-year defense plan of record. Of course those numbers are uncertain in the outyears, but they show a rise, a significant increase in fiscal year 1994. Those budget numbers have been programmed by the submarine platform sponsor the Assistant Chief of Naval Operations for Undersea Warfare.

In addition to that, in fiscal year 1990, the advanced submarine ASW development line, which is specifically targeted at improved submarine sonar systems development, contain $16.3 million, and the submarine tactical warfare systems line, which deals with submarine Arctic warfare development, has another $7.5 million.

The total comes up to in excess of $50 million in 6.3 R&D money. That is transition money, as we understand it. That is in fiscal year 1990, sir. Does that answer your question?

Mr. McCURDY. Yes, more than I probably asked for.

Admiral, your portion of the budget, we have got $54 million basically, 29 and 6.3 on whole mechanics and 16 plus 75. in the ASW advanced. Yours is what?

Admiral WOLKENSDORFER. The exploratory development side is of the order of $200 million. I can't give you a precise figure, but that is very close. Included in that are recent undersea development efforts, weapons developments, and essentially technology developments across the board.

I would like to point out that Admiral Fitzgerald, when he spoke earlier, said this problem had been recognized for some time, and it has been. In fact, much experimentation, sea tests and that sort of thing have continue gone on. Perhaps some of the lag we see, and what the panel saw, is due to some of the results just now coming in from some of the sea tests.

I think the whole issue of the technology development side has not been at a standstill. We have been working at it very hard. Again, the reason that you don't see it is that some of the results are just starting to come in now. In fact, they look very favorable and will be incorporated into some of the systems that we now have as well as systems of the future.

Mr. McCurdy. Mr. Chairman, and again, this is more the statement than the question, I appreciate the committee's indulgence. The fact of the matter is that DARPA gets about 87 million into this, and could probably use about 120. Gentlemen, is it not a fact that if the Soviet quieting threat is growing faster than our ability to keep abreast of it, doesn't the Navy believe that more than $20 million or $50 million for ASW and for submarines is justified? Isn't that just a yes or no? Isn't the threat greater than the resources that we are applying to it today?

Admiral Smith. With all due respect to your directness, I am going to give you an answer that you could consider a waffle. But let me say in the 1990 program, 5 years ago when we looked at the R&D lines that were in 62 and 63 that are now producing systems, when we look in the 63.8 and 64 lines that represent systems like BUSY-1 and BUSY-2, the update four to the advance aircraft, the B-3 and the improvements to ship sonar, at that point in time the 1990 budget for the Navy had $35 billion more in it than we have today.

During that time, there has been a growth of 25 to 35 percent in the R&D technology line. We are looking at a very small piece of the total ASW problem.

Mr. McCurdy. I understand. One last point is the fact is you are talking hardware, you are talking developments that took 10, 20 years to develop. The development cycle in our business is too damned long. If you want to get down to the bare bones, the fact is in the last 3 years, the Soviet have made tremendous progress. We have seen the classified charts and all the others and we see the degree of quieting.

We know they are making real progress. The issue is, are we keeping abreast with what they are doing? Are we developing and improving at such a rate to match that? These are the jewels of your Navy and our Navy. We are talking about the crown jewels.

Carriers are marvelous in peacetime. The submarines are the ones that are going to protect us. I believe that the threat is greater than the resources. Thank you, Mr. Chairman.

Mr. Bennett. It looks like the first priority you have in 1989 and 1990 is something you planned for some years back. The priority in getting its preeminence fairly recently has suffered by the fact that bureaucratic procedures move slowly. That is what we would like to change if we could. I think most Members of Congress feel like this is really a first priority, and we think it is being treated like it is in a mechanism that was established before it was a first priority, and it is very difficult to shake it out.

I would like to have somebody who would knock heads together and feel like he was doing what he ought to do under the format. I am not sure that is so here. I would like to change it if I could.

Mr. Hochbrueckner is next.

Mr. HOCHBRUECKNER. Thank you, Mr. Chairman.

With the new concern over the quiet Soviet submarines, it is clear that a major Navy commitment, as you indicated, is to improve your ASW capability aboard our attack submarines. The 688 is very crowded. I visited one last year. It is jam-packed with electronics, and there are very few nooks and crannies that don't have something stuffed in it.

It seems to me, as we improve our ASW capability, whether you move more into active or whatever, you are going to have to cram a lot more electronics on the existing 688 platforms, which will be around for a long time. It seems to me you have a common problem, a common concern with Naval air in that they have a very similar problem.

Just as you have to cram things into a submarine, they have to cram things into an aircraft. Of course, in your case, I would think your greater concern is more with control than it is with weight and pounds. But nevertheless, all three are tied together. So, I have two questions.

First, have you been working with the Naval air people to see what common areas you can deal in? Specifically, it seems to me you have a lot of equipment aboard ship that has been around for a while. There is nothing new and exotic. The equations have existed, the schematics are there. Do we package some of the existing electronics using more modern devices like one and quarter microns in order to cram more into a smaller space to free up room to add those other ASW systems that are going to fall out of the investment that you are making?

Admiral SMITH. That is a very good question, and a very perceptive one. In the earlier days, we solved that problem by taking out the sailor's bunk and putting the new piece of hardware in. Then we had the phenomenon called hot-bunking. You have got three sailors to two bunks. That is not the quality of life standard that the Navy would like to set for its modern ships.

In response to your question, we very much have an aggressive program going. Mr. Bob Moore talked earlier about improvement, both in the automation with the requirement for fewer sailors and more processing. I would ask Admiral Evans to comment a little on that, because I think he has some specifics that he could give you.

Admiral EVANS. Yes, sir. In the submarine arena, the 688 improved SSN 688 class submarine is pretty much at the limit of what we can put on board that platform. The systems and equipment that constitute the suit for that ship are dense—have you been aboard the San Juan? Then you have seen it first hand. That is one of the reasons why we must move forward to the new SEA WOLF class submarine, which has the design margin built into it to permit us in the future to add systems and equipment that emerge from the advanced submarine technology program that I am responsible for that will be fed by other initiatives, including sensor and combat systems, initiatives that DARPA might undertake.

Mr. HOCHBRUECKNER. I understand the point for the 21. There is no argument there. That is coming. But the fact still is that there will be 688s around for a long time, and there are going to be many new ASWs, active or passive, that need to be retrofitted into 688.

It seems to me the only way you are going to be able to do that is to take some of the existing hardware and repackage it, thus making it smaller, lighter——

Admiral EVANS. Yes, sir, I fully agree with you. I was going to continue to say that there is a synergy here in some of the initiatives that Mr. Moore described. As we move forward with the examination of light-weight materials, including metal matrix, the composites and things of that nature, that all of us are working on, both DARPA and the Navy and the National Laboratories, we are looking for materials that we can substitute for existing material structures on the 688 class to permit additional systems to be installed that would be lighter and go in the same footprint location.

We believe that VHSIC and other advanced electronics miniaturization initiatives will permit us in future electronics systems to reduce size and weight, and allow us to install those systems to gain the additional capabilities we will need to upgrade those ships.

I think we rely heavily on the defense electronics industry to proceed forward with their independent R&D dollars and their own corporate investment dollars to achieve these results. DARPA is also working on advanced computer concepts and processing techniques which we believe will add to our capability to do those sorts of things.

Mr. HOCHBRUECKNER. Let me respond by saying over the last 2 years the Congress has been promoting a program called "x-ray lithography," which will produce the next generation of computer chip to a quarter of a micron.

I might suggest that we have funded it in the Congress over the last 2 years to the tune of 25 million the year before last, 20 million in this year. We need 40 million in this year.

There is no money in the pot. DARPA didn't put money in. Here is a program that I believe is very important to your submarines and your capability. We as a Nation, I think, need it.

I might suggest if you want to make a good investigation, find a quarter of a million dollars, give it to DARPA and let's carry on with this important program.

After 2 years of putting it in, it is a vital program, clearly something you need.

With that, I will close.

I also would like to say Admiral Wolkensdorfer, I am delighted to see a man with 13 years in his name. I am partial to long names.

Thank you, Mr. Chairman.

Mr. BENNETT. Mr. Dornan.

Mr. DORNAN. We are going to go into executive session soon here?

Mr. BENNETT. Yes. I will ask in a minute we will ask whether we feel like we should.

Mr. DORNAN. I will save my questions until then, Mr. Chairman. I was called out of our hearing here a second ago on a totally different subject.

I had two constituents guests from the California State Building and Construction Trade Council.

One of them asked me what the hearing is about. It turns out he is a former submariner who served on the Catfish in the mid 1950s. I asked him how long the Catfish was. He said, 312 feet.

You mentioned, Admiral, you have a quarter size model of the Sea Wolf?

Admiral SMITH. We do have a quarter size model.

Mr. DORNAN. Doing hydro-flow tests?

Admiral SMITH. Yes, to understand the dynamics of it as it moves through the water.

Mr. DORNAN. Where is that model?

Admiral SMITH. In Idaho. We have a fresh water lake out there that allows us to test it where we can control the type of water, the temperature.

Mr. DORNAN. Have any Congressman come out and taken a look at that?

Admiral SMITH. You would be more than welcome to.

Mr. DORNAN. How long is that model, about the size of the Catfish?

Admiral EVANS. Ninety feet long, sir. It is the Large Scale Vehicle (LSV). It is robotically controlled. It is self propelled, self-controlling and is pre-programmed for its runs. It is a highly capable vehicle that is, in fact, a quarter scale model of the SSN-21.

Mr. DORNAN. How long will the program be in existence out there? I am trying to get an idea of when some of us could come out and take a look at it?

Admiral EVANS. The program will continue to support the SSN-21 program for another year and a half. Then we will transition the Large Scale Vehicle to a Navy test resource that we will be using not only to do Navy hydrodynamics, hull and propulsion research, but also it will be the test bed for a number of DARPA programs that we are currently discussing with DARPA.

Funds for that program to continue under Navy generic sponsorship are contained in my fiscal year 1991 budget, sir.

Mr. DORNAN. My constituent John Monaro mentioned to me he joined the Catfish within days after an accident in the San Francisco Harbor where during a battery charging incident there was an explosion and three soldiers were killed.

Has there ever been—I know there is operational training accidents, but has there ever been a sailor killed due to anything, in any way to do with nuclear propulsion systems?

Admiral EVANS. No.

Mr. DORNAN. That is something the Navy can be proud of.

I will have the rest of my questions for the closed session.

Thank you.

Mr. BENNETT. Mr. Rowland.

Mr. ROWLAND. Thank you.

Admiral Smith, in your testimony, you made reference to an ASW assessment conducted by the Director of Naval Operations. You stated the U.S. will have qualitative superiority and submarine platforms in the year 2000.

Would you be kind enough to elaborate a little bit?

In other words, in reference to qualitative do you,, for example, determine the damage done by the Walker spy ring? Has that been built into this analysis in determining the qualitative advantage?

Admiral SMITH. It certainly has been. We built in their technology capabilities against our projected non-abilities.

Mr. ROWLAND. We still can determine qualitative superiority in the year 2000. I am intrigued by the word "qualitative" and would go a step further and ask you if you can determine whether we will have quantitative advantage or disadvantage in that year.

I am glad my colleague Mr. Bateman referred to the 688 because it would be my assessment is that we will probably hurt our quantitative advantage if we don't proceed with the last two 688s.

Would you agree or disagree with that? I see in light of the state-of-the-art ability that the 688s have?

Admiral SMITH. I would agree with you. In the quantitative area we are looking at a Soviet submarine fleet that is listed as some 300 or 330 submarines.

They are not all nuclear or current generation. There are only a very few.

We have elected not to get into a quantitative comparison. Not that we couldn't, but we don't know what degree they will elect to keep some of those older ships in use.

The qualitative was a much more reasonable and much more defensible position to take.

In the normal under sea environment, the submarine tends to meet in a one on one situation.

Mr. ROWLAND. If we did not produce the last two 688s, would that—I think you are agreeing that would certainly affect the quantitative.

Would it also have an impact on our qualitative capabilities in the future?

Admiral SMITH. That is correct.

Each ship has the lessons learned from the prior ship. We feel each one has a very, very slight margin against the other.

Interestingly enough from a point of history, that would be the first year since 1953 that the U.S. Commerce or the U.S. Defense Department decided they couldn't afford a submarine, attack submarines in its budget.

Mr. ROWLAND. I am glad to hear you will be fighting for it. With regard to the SSN-21, there is approximately $861 million for advanced procurement.

Will that satisfy the direction of the SSN-21 at this point.

Admiral SMITH. Yes, it will.

Mr. ROWLAND. Thank you, Mr. Chairman.

Mr. BENNETT. Mr. Hunter.

Mr. HUNTER. No questions.

Mr. BENNETT. Since we are probably approaching a vote soon on the floor, everybody please leave except the witnesses and those backing them up.

Mr. KYL. Referring again to the panel, the panel concluded the ASW technology research had real growth in funding on a long term basis.

Do you agree with that? If so, what growth in ASW technology research have you planned?

Admiral SMITH. We definitely agree with that. The funding rose through 1991. I think it is premature to read too much into whatever the funding lines are beyond 1991 given the instabilities that the Defense Department and the Navy budget has had over the last couple of years.

Mr. KYL. I think I had the 1989 numbers, but the 1990 and 1991 numbers, can something give that to me?

Admiral EVANS. I am sorry, sir. I couldn't hear you.

Mr. KYL. Total ASW funding?

Admiral EVANS. I don't have those numbers, sir. I just have the submarine technology part of the equation.

Mr. KYL. We need to get those numbers for the record.

I would like to have somebody get them to me as soon as possible, if you don't have them at your fingertips.

Admiral EVANS. We will take it for the record, but we may be able to find it before the hearing ends.

Mr. KYL. Has anything the panel said or recommended in its report or the testimony that occurred on March 21, resulted either in any organizational or in any budgetary changes on your part?

Admiral SMITH. The original changes we talked about earlier were given by the Defense Science Board Report. Since the panel's report, we have, of course, been in a somewhat of a hiatus within the Department of Defense as well as the Department of the Navy's leadership.

It is turning over the rock, so to speak, as we change from one set of civilian leaders to another.

I think it wouldn't be appropriate to comment on what further organizational changes in advance of new Secretary of the Navy.

On the budgetary side, I am sure you appreciate that we really have an opportunity within the Navy to have our own control a budget once in every 2-year cycle. In the 1990 budget that opportunity was between February and April of last year.

Once we make that change, then any future changes to our program are not necessarily under Navy's control. They are directed by the combination of hat we think we can do and what OSD will permit us to do.

So there have been, there have not been any changes to the 1990, 1991 budget because that budget has not really been opened up in such a way that we could add money to those programs.

Mr. KYL. Mr. Chairman, I think in view of the question that you asked when you opened the hearing and some of the other questions that have been asked here, I would like to have a specific response to these two points.

As I understand what you are saying is, because of the change in the civilian side of the operation and because of the previous development of the budget, the answer to the question is no, the panel's testimony and recommendations have not impacted on it yet.

I think it is important for us to know very quickly to what extent the panel recommendations will impact on both the organization and the budget side, and would request that as soon as you can answer that question for us, in letter or briefing or whatever form is appropriate.

In the meantime, let us know about when you think you might be able to do that.

Would that be a fair request, Mr. Chairman?

Mr. BENNETT. I think——

Admiral SMITH. Certainly.

Mr. BENNETT. I am not quite sure. Is the room cleared now. I don't have any questions to ask.

[Whereupon, 3:40 p.m., the subcommittee proceeded to other business.]

EXECUTIVE SESSION

The subcommittees met, pursuant to other business, at 3:40 p.m., in room 2118, Rayburn House Office Building, Hon. Charles E. Bennett (chairman of the Seapower and Strategic and Critical Materials Subcommittee) presiding.

Mr. BENNETT. The hearing has come to an end. The briefing has started.

Mr. Dornan said specifically he wanted to inquire. I will ask him to inquire first. The rest of them on the list may ask questions as they come in. Mr. Dornan held up his questions.

Mr. DORNAN. Thank you, Mr. Chairman.

I just wanted to inquire about the sense of urgency that I don't necessarily see there in response to one of Mr. McCurdy's questions about he didn't see the allocation of effort, particularly in dollars, to match the evolving Soviet threat. When I go to the open press, "60 minutes", "20/20", programs like that, it seems like every time somebody touches on or writes on the Walker scandal and other spy scandals, it becomes an even greater tragedy, the enormity of classified information that appears to be lost.

I want to ask you a very simple question. Is there a sense of urgency in all of the various fields of study that you gentlemen represent and Mr. Moore, also, from DARPA, a real sense of urgency that there has been a tremendous national loss in the gap of technology between our skills and Soviet skills in submarine warfare that would develop a sense of urgency?

Do you people feel this enormous loss that we hear about on the programs because of treason?

Admiral SMITH. I would like to ask our Deputy Director of Naval Intelligence to give you some perspective on that. He is seated immediately behind me, Mr. Rich Haver.

Mr. HAVER. Everyone we talked to in the Navy feels that loss. I would say that the way we have characterized it, it isn't as though the Soviets suddenly developed submarine quieting by something Walker gave them. We have to give them credit for doing the engineering and doing the hard work. John Walker didn't have access to the Electric Boat or Newport News design people that engineer propulsion systems. Without question, we believe John Walker gave them access to the information about how we were able to exploit the deficiencies they were suffering.

If anything, that information gave the Soviets the sense of urgency, the funding, to do whatever it was necessary to correct some of their problems. I should add, they have not corrected them completely.

From all we can tell, the Soviet engineering bureau that produced the reactors and turbo generators in the Sierras over the last 15 years are still hard at work. The Soviets have not achieved the total submarine package that they are after. I think that is why you see it, Mr. Murray's committee report, suggest that there is more to come. The story is not yet completely written.

[Deleted].

If you could think of it this way: from the mid-seventies to the early eighties, the Soviets had sufficient access to our intelligence reports to see where we thought their submarine force was on a day-in, day-out basis. They also were able to go back in their own archives and get their own patrol reports and know exactly where they were.

They have a better idea of how good or bad we are in ASW than we do, because they have the ground truth. They know where they really were and they know where they think they were. From that, I think the Soviets developed an appreciation that we were good and their submarines were too vulnerable.

That is what has created the submarine force you see emerging from the Soviet shipyards today. That is what Walker and others did to us.

It isn't as though he made the U.S.S. *Queenfish,* for example, more vulnerable, but it clearly made Akula Hull 2 less vulnerable to the ASW system.

[Deleted].

Mr. KYL. One of the suggestions of the panel was better passive sensors are not the answer. There was a graph in there for why the reasons for that were so. Admiral Smith, I think one of you testified—Dr. Moore, I think you—or Mr. Moore, I think you testified to the effect one of the efforts was to develop a two-phase sonar. Is that misdirected or am I missing something here?

Mr. MOORE. I think you are absolutely right. Well, I guess we are a little more positive that we can get back capability that has been lost by this two dimensional technique.

It is not easy.

Mr. KYL. Will it be that effective given the very narrow band of improvement that would result from the development of that technology?

Mr. MOORE. We are going for much larger band of improvement than was reflected in that craft, in the report. I am not sure what specific system they were thinking of.

Perhaps the submarine sonar. I think that graph would apply to a submarine sonar, but I don't think that graph applies to a large two-dimensional array in the open ocean.

Mr. BENNETT. Anybody on the panel think they should tell us something we have not heard?

Admiral SMITH. I think Admiral Wolkensdorfer would like to say a couple of things about Mr. Kyl's question.

Admiral WOLKENSDORFER Concerning the portion we have been working on in the past, it is a very small section of the [deleted].

We feel very confident.

Mr. BATEMAN. Mr. Chairman, I don't know whether any response to this need be in executive session or not, but in case it is, I would like to ask it now.

I have sat here through this hearing as I sat through most of the previous hearing with the panel of experts. What seems to be involved as it comes through to me is, is the Navy, is it Department of Defense, or is the country putting enough priority on the challenges of anti-submarine warfare?

The Navy tells us, yes, indeed, we do recognize the priority it should have, and we are acting accordingly. I understand Mr. Moore to be telling us essentially that DARPA understands the problem and is giving it the requisite priority.

My colleague, Mr. McCurdy, on the other hand points out to how incredibly sensitive it is to our national security that we be on top and superior in this area and maintain the technological superiority we have enjoyed in past years.

But how can you say this is in effect Mr. McCurdy's point of view, that you recognize the appropriate priority for this because you aren't spending enough money on it.

I hear you saying, we understand this to be a priority of highest order for our national security. That you are doing everything you know how to do and you are exploiting all the technology you know how to exploit or that is available to you or that looks promising. I am a little uncomfortable when we measure national priority in terms of how much money Congress is appropriating for a specific area of inquiry or endeavor.

Some things of inordinate priority don't cost as much money to solve as some things with less priority, but, if you are going to do anything, still require giant sums of money. Do you perceive that we members of the committee are trying to equate situations of the national defense priority purely to number of dollars put in a budget as it works through the Congressional legislative process?

Admiral SMITH. Not at all. I think the committee has been most supportive, most forthcoming on our issues, and they are truly, those interests on your part, are truly and genuinely appreciated.

Let me speak to the issue a minute.

Mr. BATEMAN. That is the issue I want you to speak to. Are you asking for enough money? Are we giving you enough money?

Are we at risk, in your view, of us throwing money at you that will be improvidently spent?

Admiral SMITH. The question we come to in the overall allocation—one of the reasons the C&O asked me to come lead this group for today's hearing—is to make sure it was well understood that there could always be more money spent, but we currently are spending every dollar wisely and effectively within the ability to manage that dollar. We have enough good people. We have the best talent that Navy has in civilians and in military working the ASW problem.

It is not just the crews on the submarines, it is not just the people in the aircraft. It is the support of those people in the industries and in the laboratories. The Navy laboratory system is the best advanced development organization within the Defense Department and is recognized as such.

Those people give ASW a top priority. Could we spend more money? The answer is yes.

Can we spend it more effectively?

The answer is maybe.

Will there be spillover and less use of those dollars?

The answer is yes. When we made the tradeoffs developing the 1990 budget the tradeoffs indicated we could spend these dollars with a very high return on the investment.

The problem is that this is a once every 2-year window for us. Once the Secretary's staff gets involved, it is not just our priority, and then we have to justify, and rightly so, why we have changed our priorities in the middle of the process. So we feel that we are getting every dollar's worth out of it and the panel did not come before the committee today to say that we cannot do the current job with the urgency we put on it unless we have more money.

Mr. BENNETT. There is a very practical question I would like to put to you. It is based on the fact that there is no way in which very substantial sums and some operations in the DOD budget are going to stay at the high levels they are at now. There are going to be things reduced.

If I had my druthers, the SDI would be reduced. That is my view. But there isn't any doubt there are going to be some cuts. When the cuts come, I have never tried to cut just to cut SDI or something like that. I cut with the idea of strengthening our country with things that are more important.

Therefore, the question I am asking you right now, you may not be able to do this at the moment, but I would like you to give to this committee, if you can, the things that you really feel—and I ask DARPA to do the same thing—the things that if you had the money to do something in ASW that you are not now doing, what would be the amount and what is the program you would like to operate on?

It is very real that there will be reductions in the budget. You wouldn't have anything to do with it. The fact that you gave these opinions to us wouldn't have anything to do with it at all.

We will be able to cut some figures in the budget. When they are cut, I would like to ascribe them to things that I think are really important to our national defense.

There is no particular thing in the budget this year that has got the attention more for myself, and I think most Members of Congress, than ASW and what we can do about it.

We don't want to throw money away. If there is something we can do with a certain amount of money, let's do it. We are going to have some cuts. We are going to meet within the budget, but things are going to be rearranged to some degree.

Maybe SDI, maybe something else, but there will be something cut. I would really like to have you furnish for the record where you would like to do it if you need some breakthrough in ASW.

Mr. BATEMAN. Mr. Chairman, I want to thank you for asking me to yield the time because I think you are right on the point that needs to be made here. Everyone concedes what a very, very high priority this is.

Tell us what you can't do for lack of money so we can see that priority is reflected. I don't want to be in a position of just saying, as a committee, "Here is another $100 million, go out and spend it because we think this is an important problem."

Show me something not being funded that will help us solve this problem to be better at ASW, then certainly this Member is going to want to support it.

Mr. HERTEL. I want to concur with the last two gentlemen.

Mr. BENNETT. Mr. Hertel.

Mr. HERTEL. Clearly with the situation as it is today and some of the problems in the other two legs of the triad, it becomes more important and the enormous investment our Nation has made, we understand the frugal times that we are in. When you are going to be getting less money in 1989 than 1988, substantially less, and the Chairman points out we are going to cut further, we are concerned that maybe you won't have enough.

We are happy when people come and say they are doing the best they can with the money they have, but we are concerned about the future and this is not a program we want to underfund.

Could you address the less money in 1989 than 1988, how that will affect you?

Admiral SMITH. Clearly in spite of whatever the cuts will be, the production will go as first priority to the ASW R&D programs. That has been reaffirmed and at the present time, there are no ASW programs that are in R&D that are not important to us. None of those programs are being looked at for reductions by the Secretary of Defense. He is supporting our ASW initiatives as you currently see them in the budget.

Mr. HERTEL. The much larger reduction is projected for 1990 in the DARPA budget for ASW.

Admiral SMITH. I don't believe it involves ASW R&D.

Mr. BENNETT. Are you telling me that this is really nothing that you are prepared—in other words, if I want to distribute some of this money that I am able to cut out of unneeded programs, I might as well give it to the Army because you can't use it in ASW? I am an Army man.

Mr. HERTEL. The report I am talking about is in response to a House request received in February, 1989, from DARPA, classified secret, high pay-off of ASW technology initiative report to Congress. That is where they talk about substantial cuts in 1990.

Mr. MOORE. That report calls for $120 million in fiscal year 1990. The amount in the President's budget request is $87 million.

Mr. HERTEL. That is the line I am talking about, yes, sir. We are already talking about in 1988, $113 million, $95 million in 1989, and going down to $87 million. There is no other area where we have a threat to our defense that deals with working to ensure our future where we are going down in real dollars that I know of in the entire Defense Department except this program. Certainly not to this percentage of something as important to our security.

As hard as people work, having less money around here doesn't seem to show either the priority or the need. That is why the House committee asked for this report specifically.

Can you address that? How can we keep accomplishing what we need to in this most Important area that we have all agreed on if we are going to keep cutting the funding every year?

Mr. MOORE. I guess the answer is, we can't. It is clearly an important area. The Defense Science Board has pointed that out, the advisory panel to this committee has pointed that out, and in order to respond to that, we need to exploit our technology, to move out with some initiatives and do some difficult technology advances to demonstrate new capabilities at sea with prototyping and so forth, and that is simply beyond our current budget capabilities.

Mr. HERTEL. But in the scale of things, if we spent 20 more million dollars for something this important and the scale of our total defense commitment in dollars, we are talking about a very small investment.

I am sorry to say $20 million gets spilled around here. When you get to conference committee, it gets very messy on the floor. We are concerned that we go into conference with figures and they have to come up with an overall number and in the wee hours of the morning sometimes things aren't weighted as they should be, and we see cuts rather than the specificity and for the responsibilities deserve.

Admiral, could you give us a list of what additional measures and areas that you would like to work in if we were able to increase the ASW spending for $20 million for 1989?

Admiral SMITH. We could.

Mr. HERTEL. That is at our request. We understand the decision of the Navy.

Admiral SMITH. Clearly, one of the things that was not addressed in the open session was that we have a substantial investment, not only of dollars, but of people, in what we call special access programs. These efforts are designed to keep the technology edge in our ASW. These programs cover a number of areas, but for now principally submarines. Those programs are well managed and highly leveraged, with the prospect of near-term payoff that will give us good confidence that we will stay ahead of the Soviets. Those programs will take us away from the narrowband analysis and the narrowband dependency that Mr. Murray's report so clearly highlights as a vulnerability.

We did not talk about that in open session for obvious reasons. Mr. Dornan, the impact of the Whitworth-Walker case has driven us to that. I would ask your indulgence to allow each of my associates to make a short closing remark. I think they have watched this with interest and are impressed by the interest of your committee and the membership.

Mr. HERTEL. Before that, could I enlarge my request to say, tell us what you could do if you had $20 million additional, $50 million additional or $90 million additional?

Mr. BENNETT. That is the sort of information we need. We thank every one of you for your presentation. I hope you will convey to the CNO and the Secretary of the Navy that there is more than the usual concern on the part of the committee.

We desire to see ASW move forward more rapidly. Sometimes we can save money out of one program and I have difficulty getting prompt answers. You have stimulated me to think we need more in ASW, so I am looking for you to come up with something you think is worthwhile.

If we are going to find some money, let's use it for something we really need.

Admiral WOLKENSDORFER. Sir, I would be most pleased to participate in looking at the technology side to look for things that need help and obviously I am sure that we can come up with those.

Thank you.

Admiral EVANS. Mr. Chairman, I know from the fact that the initiative initiated from this committee that the Navy was criticized on the conduct of advanced submarine research and development beyond the SSN-21, which prompted the initiative to place the program that is in place with DARPA. My office was established by the Secretary of the Navy and the Chief of Naval Operations in August of last year. Since that time, I have had outstanding support from all echelons within the Navy, from the Defense Research Project Agency and the Permanent Submarine Technology Office at DARPA, and from the national laboratories that we have drawn into our process.

I can assure you that we are focusing on the future, that we have in place the organizations and a strong process to ensure that we explore all areas of submarine technology including those being pursued by DARPA, that we will select and mature the most promising high payoff technologies for future submarine design, that we will fuse our technology development plans in each area into an integrated advanced submarine R&D plan that will be an investment map for the future and that we will be ready to initiate the full-scale development of future designs for submarines with a full range of leading edge technologies.

I appreciate the opportunity to appear today.

Thank you.

Mr. BENNETT. Thank you very much.

We will adjourn.

(Whereupon, at 4:08 p.m., the subcommittees adjourned.)

[The following information was received for the record:]

REPORT

OF THE

ADVISORY PANEL ON SUBMARINE AND ANTISUBMARINE WARFARE

TO THE

HOUSE ARMED SERVICES SUBCOMMITTEES

ON

RESEARCH AND DEVELOPMENT

AND

SEAPOWER AND STRATEGIC AND CRITICAL MATERIALS

MARCH 21, 1989

(UNCLASSIFIED EDITION)

MEMBERS OF THE PANEL:

VADM Edward A. Burkhalter, USN (RET)
Dr. John S. Foster, Jr.
Dr. George H. Heilmeier
Mr. Harry A. Jackson, P.E.
Dr. Paul G. Kaminski
Dr. William J. Perry
Dr. Harold Rosenbaum
Dr. Harold P. Smith, Jr.
RADM Robert H. Wertheim, USN (RET)
Dr. Lowell Wood

EXECUTIVE SUMMARY

Aside from the deterrence of thermonuclear war itself, our commitment to aid in the defense not only of Western Europe, but also areas such as Northeast Asia and the Middle East, has long been the centerpiece of United States national security policy. The expense of the conventional forces that that commitment entails accounts for most of the defense budget.

To honor that commitment if challenged, we would have to move our forces overseas, perhaps to remote areas, and then resupply them with the materiel of war. Because we could not build enough airlift to move more than a tiny fraction of the tonnage involved, most of it would have to go by sea.

Thus, a major part of our long-standing national security policy requires that our sea lines of communication be secure in time of war. Air attacks are one possible means of interdicting them. This report deals with the other: attack by submarine.

Since the end of World War II, the United States has built the best submarines and had the best anti-submarine capability in the world. While our military shipping would no doubt suffer losses to submarines, we are confident that we could still reinforce and support Europe or other vital areas adequately if war were to break out today.

But whether we will still be able to do that in the future is becoming much less certain. And thus our future ability to carry out a major part of the national security policy that we have pursued for the past 40 years is becoming much less certain. Obviously, this is an issue of profound importance to the security of the United States.

Our current anti-submarine warfare (ASW) capability rests almost entirely on listening for the sounds generated by Soviet submarines. That approach has been successful and our antisubmarine forces have grown potent because the Soviets have traditionally built relatively noisy submarines. But the future of that approach to ASW is now very much in doubt because the Soviet Union has begun to produce quiet submarines.

Because the Soviets cannot build enough new submarines overnight to constitute a serious threat, we still have time to do something about the problem. But do something we must: we must build what will amount to an entire new ASW capability by the time the Soviet Union has built a significant number of new submarines.

The Navy is responding competently in researching technological alternatives to our current systems. But the Navy establishment -- like many organizations of comparable size and age -- is burdened with internal vested and sometimes conflicting interests that encumber innovation and execution on the scale required here. Business as usual is not up to an undertaking of this magnitude and importance. Dramatic new initiatives are essential if the problem is to be solved in time.

We believe that the Navy must, in effect, "start over" with new approaches to ASW. This is no ordinary challenge; it is of a different nature and on a different scale than most. What is needed is not simply more money and harder work, important as those may be; what is needed is an entirely new and aggressive architecture for coping with this immensely serious development.

The Congress should make sure that that happens and that the Navy raises the priority accorded ASW. It should encourage more vigorous development of our submarine technology, not just for the new SSN-21, but for future classes too.

The Defense Advanced Research Projects Agency (DARPA) is also working on both ASW and submarine technology. Because a diversity of approaches is prudent when the technological challenge is as formidable and the stakes are as high as they are in this case, DARPA's work should continue in parallel with the Navy's, but with DARPA -- in keeping with its traditional mission -- concentrating on the more advanced and riskier technological alternatives. But though independently managed, there needs to be better communication and coordination between the two programs.

The importance of this research justifies significant real growth in funding in spite of today's downward pressure on the defense budget. Indeed, with the maintenance of our ASW capability arguably the most important of all the challenges facing the DoD today, a case can be made that funding for ASW research should be determined less on customary grounds of compromise among many projects competing for limited resources than on grounds of what could usefully and efficiently be applied to this most critical of all R&D tasks. But whatever the level, the funding should be predictable, enduring, and stable.

By Congressional direction, the CIA has undertaken a novel role: basic ASW research and experimentation at sea. We believe that funding for the program should continue, but with a gradual shift away from basic research and back toward the Agency's fundamental mission of intelligence as other independent organizations become able to respond to the research and experimentation needs of the Agency. But to understand what data to look for, how to obtain it, and how to interpret it, the Agency must retain the scientific expertise that it has built over the past.four years. It can do that by to continuing to participate in the planning and analysis of submarine and ASW research.

And finally, because of the critical importance of the issue and because of the number of technical, intelligence, and operational communities involved, we recommend that the Secretary of Defense appoint a broadly based standing committee of recognized experts in the relevant disciplines to advise him on the subject of ASW, including:

o   The evolution of the submarine threat;
o   Objectives, strategy, plans, and programs to deal with that threat;
o   The adequacy of U.S. programs, management, and resources;
o   The quality of U.S. research;
o   The adequacy of coordination among the organizations involved; and
o   Desirable actions and initiatives.

Further discussion of the issues in this summary is in the appendix.

APPENDIX TO THE REPORT
OF THE
ADVISORY PANEL ON SUBMARINE AND ANTISUBMARINE WARFARE
TO THE
HOUSE ARMED SERVICES SUBCOMMITTEES
ON
RESEARCH AND DEVELOPMENT
AND
SEAPOWER AND STRATEGIC AND CRITICAL MATERIALS

SUBMARINE TECHNOLOGY

The initial motivation in forming this panel was a concern that Soviet submarine technology might be outstripping ours. At the time, antisubmarine warfare (ASW) technology, though also of concern, was of somewhat secondary interest.

But as our exploration progressed, it became increasingly clear that of the two issues, ASW was not only the more pressing but, indeed, a matter meriting the highest priority of attention. The bulk of this report accordingly centers on that issue, but that is not to dismiss submarine technology.

As to the original concern that Soviet submarine technology might be outstripping ours, it is true that the Soviets' submarine R&D program is extremely ambitious, seems to overlook no promising technologies, and -- in that it dates back for many years -- is no flash in the pan. As a result of their years of intensive research, it appears that the Soviets may well be ahead of us in certain technologies, such as titanium structures and control of the hydrodynamic flow around a submarine.

But far more important is the improvement that the Soviets have made in submarine quieting. The problem is not that Soviet submarines are now quieter than ours; they are not. But after decades of building comparatively noisy submarines, the Soviets have now begun to build submarines that are quiet enough to present for us a major technological challenge with profound national security implications.

Though we see no evidence as yet that the Soviets' ambitious R&D program has produced a submarine that is qualitatively superior to ours, their continuing, ambitious technology program may well produce such a submarine unless our own R&D efforts at least match theirs in scope and productivity.

It is true that, except in the case of direct submarine-to-submarine encounters, how well one side's submarines happen to stack up against the other side's submarines is much less important than how well they stack up against the other side's ASW capabilities. Nonetheless, submarine-to-submarine comparisons still have important implications for other than submarine-versus-submarine operations: a Soviet superiority would imply that we had not taken full advantage of what the Soviets had shown to be possible, and that the effectiveness of our submarines was therefore less than it could have been.

In our opinion, the Navy's current submarine technology program is unduly restricted to issues relating to the design of its forthcoming class of attack submarines, the SSN-21 *Seawolf*. While the importance of technological excellence in this new design is beyond question, we believe that more effort on research and development applicable to later designs is equally warranted.

The most fundamental issue in considering future SSNs is what their missions should be. The advent of very quiet Soviet submarines (as discussed below) threatens what has traditionally been the most important wartime mission of our SSN force: hunting enemy submarines. The possibility of mitigating that degradation through new technology and tactics needs to be considered now as we explore alternative designs for new SSNs. In addition, we need to consider other roles for submarines which, with their inherent stealth, can penetrate areas denied to surface ships and aircraft.

But whatever the future missions of SSNs may turn out to be, the opportunities for technological progress are rich and merit serious investigation now. We caution, however, that the objective must not be mere technological virtuosity, but real effectiveness at sea. For example, improvements in speed and depth capability, while possibly dramatic, might turn out to do less for combat effectiveness than an equal investment (dollars, space, or weight) in other kinds of improvements.

To one degree or another, work is indeed going on in such areas, both in the Navy and in the Defense Advanced Research Projects Agency. To the Navy's credit, it has decided to take an operational submarine out of the line and make it available for research and development projects. Though not on a scale with the revolutionary submarine *Albacore* built early '50s entirely for R&D purposes, the panel welcomes this initiative as very much a step in the right direction. But, nevertheless, it is our opinion that the Navy's submarine technology effort is still too narrowly concentrated on the near term, i.e., the SSN-21, to the neglect of even greater longer-term improvements. And, given the immense benefits that stemmed from *Albacore* and the growing importance of submarines, we believe that the Navy should resume experimental submarine programs for research into advanced technology.

The Navy's rationale for its emphasis on the shorter term is that it is the more pressing problem, and that full funding for both simply isn't available. But we believe that that is too narrow a view. First, much useful theoretical and experimental work can be accomplished at moderate cost. But second and more importantly, the premise that additional funding is not available assumes that the present general allocation of funds among all Navy claimants is somehow immutable. In terms of the authority to reallocate requests for funds within the Navy, of course it is not; the Navy leadership obviously has such authority. But in terms of politics within the Navy, the premise may be closer to the truth. Those who might suffer in a reallocation of funds to longer-term submarine R&D at their expense have apparently been successful in more-or-less holding the *status quo*.

We therefore recommend a program of research and development on submarine technology with the following characteristics:

1) Research on advanced submarine technology should continue in both the Navy and DARPA. The two should be independent in approach, but not in planning. Though both should be aggressive, DARPA's program should concentrate on the more venturesome projects involving higher risk but greater potential benefit.

2) The initial funding level should be no less than that of FY89, plus enough to support additional research by the Navy on submarine technology issues beyond those now being addressed in support of the SSN-21 development program. We emphasize, however, that the level at which these programs are funded is no more important than how well they are conceived and managed.

3) For both efficiency and productivity, funding for both of these programs should be established on a predictable, enduring, and stable long-term basis. Growth is advisable in a program with such rich possibilities and of such national importance, but growth is no more important than predictability.

We also note in closing that, in our opinion, the Navy's "in-house" technical community -- its laboratories and its shipyards -- has become too oriented toward maintenance of the existing fleet at the expense of research and development for the fleet of the future. For example, we doubt that it would be possible today to create the equivalent of the Special Projects Office that developed the Polaris system because we no longer have the technically trained officers experienced in ship design, shipbuilding, weapon development, and electronics who served then as the Special Projects Office department heads. We believe that the Navy should reestablish an "in-house" cadre of such technically proficient officers and civil servants.

ANTISUBMARINE WARFARE

The Decline of Passive Sonar

For decades, our principal system for detecting and tracking Soviet submarines has been passive sonar. The particular application that we and other nations have used depends on underwater microphones to detect the distinctive sounds emitted by a submarine's engine, propeller, turbogenerators, and other sources.

Our passive sonars have been able to detect the typically noisy submarines that the Soviets have built over this period at relatively long ranges. But the era of relatively noisy submarines is beginning to draw to a close. Not only are the Soviets finally beginning to build quiet submarines, such as the *Akula* class, but significant innovation is taking place in some Italian, German, and Swedish designs for non-nuclear submarines.

The traditional non-nuclear submarine, with diesel-electric propulsion, can be extremely quiet when running submerged on its batteries, but can neither stay submerged for long nor move very fast. However, new forms of non-nuclear propulsion -- closed-cycle diesels, Stirling engines, fuel cells, etc. -- are showing the potential for vastly greater submerged endurance than a conventional diesel-electric boat, while retaining its low acoustic signature, and still at a fraction of the cost of a nuclear submarine.

Being the product of foreign technology, such submarines could well become available to third-world nations in the next decade. That is a matter of special concern in that our carrier battle groups often operate near -- and from time to time in the recent past, even against -- such nations. That raises the possibility of small but radical powers being able to threaten the capital ships of the U.S. Navy. And beyond that sort of contingency, the prospect of even one such stealthy submarine under, for example, the Libyan flag entering New York harbor undetected and carrying out its mission is a matter of grave concern.

Thus with the advent of quiet Soviet nuclear submarines and the prospect of even quieter non-nuclear submarines with considerable submerged endurance but of indeterminate nationality, the effectiveness of the passive sonar systems on which we have come to rely so heavily is now being threatened. Thus we are beginning to lose the traditional mainstay of our ASW capability.

The following schematic graph, plotting the level of the sound of a submarine versus its range from our passive sonar, illustrates the underlying problem:

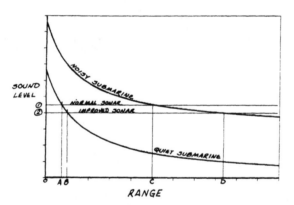

The two curves show how the sound of the submarine becomes fainter as it radiates out through the ocean. The upper curve ("noisy submarine") represents the kind of Soviet submarine the Navy has been dealing with in the past; the lower curve ("quiet submarine"), what it will have to deal with in the future.

Both curves are steep at short ranges because, in simplest terms, the

4

sound, expanding spherically through the ocean, diminishes with the square of distance (doubling the distance cuts the sound by a factor of four). The curves flatten out, however, when the sound, having reached the ocean bottom, thereafter expands cylindrically rather than spherically, and thus diminishes directly with distance rather than by its square (doubling the distance cuts the sound by a factor of only two rather than four).

The horizontal lines drawn across the graph represent the faintest sounds that can still be detected by two passive sonars, one labeled "normal" and the other, which can detect somewhat fainter signals, "improved".

The sound emitted by the noisy submarine, after starting high on the graph, falls away until, at the range labeled "C" on the graph, it can still just be detected by the "normal" sonar. The "improved" sonar, on the other hand, can still detect the noisy submarine much farther away -- all the way out to the range labeled "D".

The noise emitted by the quiet submarine starts much lower on the graph, and quickly drops to the faintest levels that can be heard by either of the two passive sonars: at range "A" for the normal sonar and range "B" -- not much farther out -- for the improved sonar.

The characteristic shape of the curves -- initially steep, but flattening out with increasing distance from the submarine -- leads to two critically important conclusions:

o   By quieting their submarines, the Soviets are drastically reducing the detection range of our passive sonars: on the graph, from "C" back to "A", or from "D" back to "B". By so doing, they are vastly increasing the number of passive sonars that we would need to cover the same amount of ocean (a 10:1 reduction in range would require a hundred-fold increase in sheer numbers of sonars for equivalent coverage).

o   Better passive sonars are not the answer. As can be seen from the graph, though the "Improved Sonar" does have a much greater range than the "Normal Sonar" against the old noisy submarine (on the graph, "D" instead of "C"), because of the steepness of the curve at short ranges, there's very little difference against the new quiet submarine ("B" instead of "A"). The improvement in sonar sensitivity that would be needed to restore the detection range to what it was against noisy submarines is simply not in the cards.

Our best current ASW systems all rely heavily on passive sonar: SOSUS (Sound System for Underwater Surveillance -- a network of large, fixed arrays on the ocean floor), SURTASS (Surveillance Towed Array Sonar System -- long arrays towed by surface ships), hull-mounted or towed sonar used on surface warships and submarines, and expendable buoys used on aircraft.

Thus, the loss in effectiveness of passive sonar due to the quieting of Soviet submarines will affect virtually every phase of our ASW capability. And a loss of ASW effectiveness raises profound national security problems such as

our ability to reinforce Europe by sea in time of war, the survivability of our
carrier task forces, and our ability to secure our coasts from missile-firing
submarines. Furthermore, because many of those wartime problems stem from
capabilities that are demonstrable in peacetime, they present opportunities for
Soviet coercion. That being the case, alternatives to passive sonar, at least
as we have used it in the past, have now become a matter of profound
importance.

Dealing With the Coming Decline in Passive Sonar

One alternative to passive sonar as we have known it -- a limited number
of relatively long-ranged passive sensors -- is a larger number of short-ranged
passive sensors in a closely spaced network. While such systems could be useful
at chokepoints such as straits that Soviet submarines would have to transit
before they could threaten us, the number of sensors required to cover the open
oceans would be prohibitively costly.

The most extensively used adjunct to passive sonar has been active sonar,
known prior to WWII as "ASDIC". Instead of listening for sounds emitted by the
target submarine, active sonar generates its own pulses of sound and listens
for their echoes reflected back from the target submarine. Thus, it -- unlike
passive sonar -- can measure the distance and closing speed of the target. But
its range has been much less than passive sonar's, at least against noisy
Soviet submarines. And, unlike passive sonar, its loud "pings" alert the hunted
to the presence of the hunter. Nonetheless, because it is not affected by
Soviet quieting, active sonar is receiving new-found attention.

Non-acoustic techniques are also possible. One long in use has been
Magnetic Anomaly Detection, or "MAD", in which a sensitive magnetometer,
usually carried by an aircraft, detects the very slight distortion of the
Earth's magnetic field caused by the steel hull of a submarine. Its range,
however, is so limited -- only a fraction of a mile -- that MAD is, and is
likely to continue to be, useful only for the final localization and confirma-
tion of a contact made by other means.

Many other non-acoustic techniques are at least theoretically possible.
Both the United States and the Soviet Union are exploring such alternatives or
adjuncts to passive sonar. It is far too early to judge the success of these
efforts, particularly because much of this work is pressing the outer limits of
science and technology -- from an understanding of the underlying physics of
the various phenomena, all the way to highly advanced sensors and data processing equipment and techniques. Apart from the Strategic Defense Initiative, this
work is probably the greatest technological challenge facing the Department of
Defense.

The Navy's Reaction to the Coming Change

The introduction of quiet Soviet submarines could well lead to a fundamental change for us in naval warfare. Unless ways to compensate for the
coming degradation in passive sonar can be found:

o   Our attack submarines are likely to lose much of their effectiveness
     in what has been one of their major missions -- operations against
     Soviet submarines;

o   Because they will be so much harder to defend against, Soviet attack
     submarines are likely to become far more effective in operations
     against our surface ships, both naval and merchant; and

o   Deprived of cuing, both our surface warships and our aircraft could
     have great difficulty in open-ocean ASW.

As noted above, these developments -- which seem likely in the absence of
adequate replacements for our traditional passive sonar systems -- threaten the
survivability of our carrier task forces, our ability to deploy and support our
military forces overseas, and the security of our coasts against submarine-
launched missiles. The Panel believes that while the Navy appreciates the
gravity of these developments, the scale of its reaction to date does not seem
commensurate with the scale of the coming challenge.

We do not mean to imply that the Navy has totally failed to react to these
developments. Indeed, it has a number of programs under way. Rather, it is both
the scale of the new initiatives and the Navy's management of change that
troubles us. The Navy's planning, programming, and budgeting system has long
been dominated by so-called "platform sponsors" representing the Navy's three
major components: the three-star Assistant Chiefs of Naval Operations for
Undersea, Surface, and Air Warfare. We believe that each of these components,
concerned as to how the coming changes in naval warfare might affect their own
interests, has been reluctant to endorse possibly expensive new initiatives,
and has instead tended toward conservatism and the *status quo.*

These narrower, more parochial interests are supposed to have been -- and,
no doubt, to some extent have been -- tempered by offices of equal rank but
with a broader scope of responsibility: the Deputy Chiefs of Naval Operations
(DCNOs) for Naval Warfare and for Navy Program Planning. Nonetheless, the
Defense Science Board found last December that the organization was not up to
the task of coping with changes of the magnitude now in prospect.

The Panel concurs with the Defense Science Board's conclusion that the
Navy's organization has inhibited appropriate action to deal with the immensely
important challenges of the future. We, too, believe that some form of organizational change is essential.

The chief characteristic of the necessary change should be a realignment
of authority over the Navy program and budget. By "authority" we do not mean
any cosmetic reconfiguration of the Navy's organization chart; we mean the
political power that actually shapes the Navy program and budget, no matter how
the organization chart happens to be drawn at the moment.

That authority must not be the captive of the platform sponsors. It should
be able to reallocate in the Navy's annual request as much funding as is
necessary to ensure that the effort devoted to coping with the coming change in

naval warfare is at a scale appropriate to the immense significance of that
change. That, in turn, implies control over the Navy's recommended allocation
of funding among R&D, procurement, operations, and personnel, and, implicitly,
control over a major part of the future force structure of the Navy.

The change need not necessarily be *de jure*; an equally satisfactory and
probably quicker solution would be a *de facto* "reorganization" in which a) the
dominance that the platform sponsors have so long exerted over the allocation
of Navy funding were suppressed, and b) the DCNOs for Naval Warfare and for
Navy Program Planning, the Chief of Naval Operations, and the Navy Secretariat
were able to shape a program commensurate to the looming problem.

After months of struggling with the issue raised by the DSB, the Navy
leadership has proposed a plan for reorganization. Though it would create a
"czar" to manage the Navy's ASW R&D efforts, the position will be only at the
two-star level and thus still inferior to the dominant platform sponsors. We
cannot tell at this point what effect, if any, this reorganization will have,
but it seems at first glance to be embroidering the organization chart rather
than attacking the root problem of why the Navy's most serious challenge is of
a lower priority than business-as-usual issues such as force structure.

We believe that the Congress can and should measure the success of this
organizational change by two criteria:

o   Whether subsequent changes in the allocation of the Navy budget duly
    reflect the importance of the challenge of quiet Soviet submarines,
    and

o   Whether the architecture of the R&D program represents a comprehen-
    sive new approach based on a credible strategy with specific objec-
    tives, plans, and programs to deal with the problem, or whether it
    representes only a continuation of business as usual.

The first criterion above -- improved funding -- is a necessary but not
sufficient condition for success. We believe that the second criterion -- a
wholly new architecture for our research and development program -- must also
be met if we are to be successful in coping with the challenge. Such an
architecture must define a rational and coordinated approach to the problem as
a whole, defining responsibilities, funding, and schedules. All the diverse
efforts in ASW must be integrated into one logical and efficient program that
will lead in a fast-paced but orderly way from research to development to the
delivery of a new ASW capability before the Soviets build enough quiet new
submarines to defeat our existing ASW forces.

We therefore recommend that the Congress closely monitor the product of
the reorganization of ASW responsibilities within the Navy in terms of 1)
changes in the Navy budget and program and 2) development of a new architecture
for a comprehensive program to deal with the challenge of quiet Soviet submarines.

DIVERSITY OF RESEARCH AND DEVELOPMENT ORGANIZATIONS

As noted above, much of the effort to develop new submarine detection systems to compensate for the degradation in passive sonar is pressing our current limits of science and technology. Furthermore, much of the effort to expand those limits is also highly classified, thus raising a problem regarding a critical element of all proper scientific investigation: publication, peer review, and replication of experimental results.

Though there is no evidence that it has happened in this instance, we believe that such circumstances can lead to problems of overcommitment to certain avenues of exploration to the neglect of others, overconfidence in their merit, and insufficiently disinterested interpretation of test results. We believe that the best insurance against such risks is a diversity of research and development organizations.

Such diversity in the form of two major undersea warfare research efforts outside the Navy has already been established by direction of the Congress. One of these is at the Defense Advanced Research Projects Agency, the other at the Central Intelligence Agency. Because the ultimate responsibility for ASW must remain with the Navy, however, these independent activities must be considered as supplements to, and not replacements for, the Navy's own ASW R&D program.

The principal value of such independent research is realized only when it differs with and challenges the Navy's own research on valid grounds. But the independent investigators' willingness to play such a role depends on their reasonable confidence in their own job security. To the extent that they fear that the Congress might allow their funding to lapse, they may be less than enthusiastic about taking prominently independent positions that might alienate the Navy, their most likely alternative employer in this highly specialized field. Thus, if the Congress endorses the concept of independent research, as we do, we recommend that it indicate a strong commitment to continued funding in the future, subject of course to reasonable performance.

The participation of the CIA in a program not only of intelligence, but also of basic ASW research and experimentation merits additional comment. A more intimate understanding of the relevant physics is clearly of value in targeting and interpreting intelligence on Soviet activities. And the quality of the Agency's research is a tribute to the individuals who have managed it. Nonetheless, it is true that running a program of basic research and experimentation is a marked departure from the Agency's traditional pursuits.

To carry out its intelligence responsibilities, we believe that the CIA must maintain a high degree of technical expertise in undersea warfare. But rather than conducting its own independent program of basic research and experimentation as it now does, it should instead have access to and draw on the research of others, and participate in the planning and analysis of that research. Such an arrangement would enhance the Agency's ability to evaluate the gamut of research projects and benefit from the best, as well as eliminating the natural tendency to favor, even unconsciously, one's own research over that of others.

But we also believe that none of its current basic research program should
be transferred away from the CIA until a suitable independent organization is
found, not only to continue the work, but also to be available to respond to
any basic research needs of the CIA that are not being met elsewhere in the
community. And in any event, this transition should be a gradual one, carried
out over the space of several years with due consideration to the issue noted
above: the morale and confidence -- and thus the independence -- of the
scientists involved.

A STANDING COMMITTEE ON ASW

Because:

o   Even though the Navy obviously bears the prime responsibility for
     solving the problem posed by the introduction of quiet Soviet nuclear
     submarines (and potentially advanced forms of non-nuclear sub-
     marines), there is such a diversity of technical and intelligence
     organizations that can and should contribute to the effort, and

o   The challenge of finding that solution is so formidable, and

o   Success in this effort is of such importance to the national securi-
     ty,

we believe that it would be prudent for the Secretary of Defense to appoint a
broadly based standing committee of recognized and, to the degree feasible,
disinterested experts in the relevant disciplines to advise him on the subject
of ASW -- technology, operations, intelligence, and ASW's role in national
security policy. The members should have access to the full range of projects
in the field, regardless of classification, uniquely suiting them to develop a
broad overview of the effort as a whole. The committee's role would be solely
advisory, its only power being that of persuasion. Among the issues it would
address would be:

o   The evolution of the submarine threat;
o   Objectives, strategy, plans, and programs to deal with that threat;
o   The adequacy of U.S. programs, management, and resources;
o   The quality of U.S. research;
o   The adequacy of coordination among the organizations involved; and
o   Actions and initiatives that the Secretary of Defense should under-
     take with regard to ASW.

We therefore recommend a program of research and development on ASW
technology with the following characteristics (many of which parallel those we
have recommended above with reference to advanced submarine technology):

1) Research on ASW technology should continue in both the Navy and DARPA,
independent in approach, but coordinated in planning. Because the consequences of failing to find a solution to the challenge presented by quiet
Soviet submarines are so serious, we recommend that this work should be

considered as one of, if not the, highest priority activities in the DoD.

2) Given such a priority and the precedence that finding long-term
solutions should take over near-term considerations such as readiness or
additions to force structure, the funding for both programs should be
determined principally on grounds of what could usefully and efficiently
be applied to the task. Both programs thus merit significant real growth
in funding but, as with the submarine technology programs, on a predictable, enduring, and stable long-term basis.

3) The CIA should continue to develop its expertise in the technology of
undersea warfare, maintain its ability to independently evaluate U.S.
research, and participate in the planning and analysis of U.S. research
that could resolve intelligence issues. But the Agency should, on a
gradual and orderly basis, phase out of independent basic research and
experimentation. The transition should be timed such that a host agency,
as independent from but coordinated with the Navy as the CIA has been, can
be phased in to continue the basic research now under the CIA with a
minimum of turbulence, and be available to the CIA thereafter to conduct
cooperative research to address intelligence issues.

In addition, we recommend that the Secretary of Defense appoint a standing
committee of experts distinguished in the fields relevant to ASW to advise him
on the subject.

If an additional $20M, $50M, or $90M were made available to the Navy for ASW Research and Development, the Navy would invest these funds as follows: (Assumption: similar amounts would be available in FY 91).

ASW TECH BASE INCREMENT

PROGRAM (Program Element)	FUNDING LEVEL 1 FY 90 / 91	FUNDING LEVEL 2 FY 90 / 91	FUNDING LEVEL 3 FY 90 / 91
DIRECT MEASUREMENT (PE 0603747N) PROGRAM	13 / 11	15 / 14	18 / 16
ADVANCED NON-ACOUSTIC ASW (PE 0603528N - 50%) (PE 0602314N - 50%)	7 / 11	7 / 11	15 / 20
HIGH GAIN INITIATIVE (PE 0602314N)		10 /18	10 / 18
ASW OCEANOGRAPHY (PE 0603785N)		8 / 13	15 / 30
deleted (PE 0603747N - 55%) (PE 0602314N - 45%)		10 / 14	14 / 16
ADVANCED TORPEDO TECHNOLOGY (PE 0602314N)			10 / 12
ASW C^{3}I (PE 0603747N)			8 / 10
TOTALS	20 / 22	50 / 70	90/ 122

Each of the programs listed is described below:

a. Direct Measurement Program. The Navy's requirement to combat *deleted* submarines has necessitated an expansion of *deleted*

is the "target

strength" term, *deleted* This is a measure of how well an active sonar signal is reflected or echoes from the target submarine's hull. Additionally,

deleted

b. _Advanced Non-Acoustic ASW_. Funds would be applied
toward accomplishment of all of the following objectives:

deleted

would continue to be
explored under the leadership of a Navy program.
Experimental, theoretical, and analytical efforts would be
brought to bear in coordination with DARPA to examine issues
determined to remain outstanding upon completion of the
present phase of investigation under _deleted_

(2) Conduct exploratory development of non-conven-
tional and advanced _deleted_
techniques. As part of this effort, a theoretical and
experimental re-examination of _deleted_ other than
classical _deleted_ would be carried
out to determine if there are signals which provide a basis
for detection at substantially larger ranges than would be
expected for _deleted_ Advanced _deleted_
sensors would be
brought to a state of maturity where they could be used in
controlled field experiments, in particular, ones associated
with characterizing the detectability of _deleted_
phenomena created _deleted_

(3) Conduct advanced technology development of
active and passive _deleted_ detectors for tactical (helo-
based) applications. Experiments, with appropriate
oceanographic truth, would be conducted which would establish
the expected effectiveness of _deleted_
detection under the range of expected environmental condi-
tions which would be encountered during ASW operations.

Real-time signal processing _deleted_ would also
be developed. In addition, supporting exploratory development and research would be carried out focusing on explaining
and exploiting _deleted_

(4) Exploratory and advanced technology development
of in-ocean _deleted_ would be pursued. Both _deleted_

concepts will
be developed and evaluated in realistic at-sea experiments.
Basic research and exploratory development would be conducted
aimed at establishing a comprehensive understanding of and
predictive capability for _deleted_ This
would take into account the _deleted_

c. <u>High Gain Initiative</u>. This program will be brought to a <u>technology-limited pace</u> leading to early development of a full scale _deleted_ in _deleted_ Recent theoretical and experimental work indicates that _deleted_

It is essential that these predictions be confirmed rapidly in order to make proper choices with regard to _deleted_
Knowledge gained in this program will also have spinoffs to _deleted_
The funding identified will accelerate the _deleted_ test program, accelerate a comprehensive end-to-end model, and provide for engineering technology and cost-reduction efforts for various components of _deleted_

d. <u>ASW Oceanography</u>. Funds will be applied to accomplish the following:

(1) Large Area Remote sensing of the Ocean Acoustic Environment - Mesoscale features, such as fronts and eddies, have been shown to have a significant impact on ASW system performance. At present, the Navy does not have _deleted_

deleted provides great promise as a real-time, cost effective data source for initialization and update of acoustic nowcast/forecast systems in regions dominated by dynamic, mesoscale ocean structures. _deleted_ provides real-time information on the _deleted_

(2) Critical Sea Tests - The CNO initiated the Critical Sea Test series in FY 87 to reduce the risk in assessing the performance of _deleted_ sonar systems and to provide inputs to system designs. This program determines D parameters, i.e., reverberation, signal propagation, target strength D _deleted_ and background clutter (false targets, differing environments, seasons, and sea states). D Critical Sea Tests have been completed; _deleted_
Funds identified above would permit a continuation of this critical D risk reduction program and would enable an additional test in FY 90 _deleted_

This effort would be included in PE 0603785N, ASW Oceanography, vice PE 0603792N, Advanced Technology Demonstration.

e. _deleted_ Although Soviet _deleted_

substantial promise still remains in _deleted_ Over the past two years, a concerted effort has been mounted to identify and describe the total content of _deleted_

For example, *deleted*

The Naval Technical Intelligence Center (NTIC)
has published a *deleted*

f. Advanced Torpedo Technology. Funds identified will
continue in the outyears the Congressionally mandated
program to enhance technology development for advanced
torpedoes (...FY 89 funding for PE 62314 was increased by
$10M). The thrust of the program is to expand the scope and
accelerate the ASW Weapon Technology Base. The program focus
is on and beyond *deleted*   and
emphasis is placed on *deleted*
concepts. The goal is to provide technology to
counter projected significant threat advances and to ensure
technological lead.

A top-down, overall ASW systems approach is being
used. The *deleted*

notional concepts are being used to focus technology development projects. Five required capabilities have been
identified, and project tasks have been grouped into
Technology Clusters to achieve these capabilities. The
Technology Clusters are: *deleted*

The program includes
conceptual analysis, simulation, and modeling, and will
utilize test vehicle philosophy to demonstrate the
feasibility of various concepts. *deleted*

g. ASW C^3I. The ASW C^3I system must provide information
management and communications capabilities adequate to meet
the threat as it evolves into the 21st Century. This means
that ASW C^3I must not only evolve *deleted*

Real time data fusion capability must be established
to integrate all source intelligence on the threat, environmental oceanographic data, and own force data. It will
require reliable, robust and survivable communications links
between *deleted*
The $C^\sim I$ network to support the ASW
Strategy must be an integral element in the entire process.

Future ASW capabilities will embody the following features:

(1) *deleted*

deleted

to collect real time sensor data over a wide ocean area. Alternatively, these fields could be associated with a system of *deleted* which would periodically visit *deleted*

(2) Central shore-based ASW contact data assimilation, interpretation, and fusion to produce a tailored, high accuracy targeting product for platforms or long-range weapons.

(3) *deleted*

from friendly platforms or remote sensors.

(4) Specific provision for two-way *deleted* using some combination of systems such as *deleted*

As the ASW architecture matures, technology opportunities relevant to the foregoing will be taken advantage of in providing the necessary C^3I capability.

○

CPSIA information can be obtained
at www.ICGtesting.com
Printed in the USA
BVHW071055190519
548699BV00001B/89/P